Optimisation Models and Methods in Energy Systems

Optimisation Models and Methods in Energy Systems

Special Issue Editor

Carlos Henggeler Antunes

MDPI • Basel • Beijing • Wuhan • Barcelona • Belgrade

MDPI

Special Issue Editor
Carlos Henggeler Antunes
University of Coimbra
Portugal

Editorial Office
MDPI
St. Alban-Anlage 66
4052 Basel, Switzerland

This is a reprint of articles from the Special Issue published online in the open access journal *Energies* (ISSN 1996-1073) from 2018 to 2019 (available at: https://www.mdpi.com/journal/energies/special_issues/Optimisation_Models_Methods_Energy_Systems)

For citation purposes, cite each article independently as indicated on the article page online and as indicated below:

LastName, A.A.; LastName, B.B.; LastName, C.C. Article Title. *Journal Name* **Year**, *Article Number*, Page Range.

ISBN 978-3-03921-118-0 (Pbk)
ISBN 978-3-03921-119-7 (PDF)

Contents

About the Special Issue Editor

Carlos Henggeler Antunes is a Full Professor of the Department of Electrical and Computer Engineering, Director of the R&D Institute INESC Coimbra, and member of the coordination committee of the Energy for Sustainability Initiative of the University of Coimbra, Portugal. He obtained his Ph.D. in Electrical Engineering (Optimization and Systems Theory) at the University of Coimbra in 1992. His research interests include multiple objective optimization, multicriteria analysis, and energy systems and policies with a focus on energy efficiency and demand response. He has participated in more than forty R&D and consulting projects in the domains of energy efficiency and decision support systems. He served as guest-editor of Special Issues in several journals on topics such as uncertainty and robustness in planning and decision support models, energy efficiency for a more sustainable world, models and methods of operations research in the energy sector, sustainable cities, challenges in the evolution of power systems to smart grids. He is Editor of the EURO *Journal on Decision Processes and Energy Policy*, and member of the editorial board of the *Journal of Energy Markets, Sustainable Cities and Society and Energies*. He co-authored the book "*Multiobjective Linear and Integer Programming*" published by Springer in April 2016.

Preface to "Optimisation Models and Methods in Energy Systems"

Challenging problems arise in all segments of energy industries—generation, transmission, distribution and consumption. Optimization models and methods play a key role in offering decision/policy makers better information to assist them in making sounder decisions at different levels, ranging from operational to strategic planning.

Energy systems and networks are increasingly complex; therefore, optimization models and methods are essential tools for the development of smart(er) networks within more integrated and sustainable energy systems, encompassing electricity, gas, district heating/cooling, etc., with pervasive deployment of information and communication technologies. Technical design, operational, economic, regulatory, social and environmental issues, among others, are at stake, requiring interdisciplinary approaches, with contributions from engineering, economics and social science fields to the definition of adequate optimization models and methods to support more informed decision processes.

Planning tasks are increasingly complex due to the unbundling of the industry value chain and the emergence of new players (e.g., aggregators) and market structures. The ongoing evolution of energy systems to smart grids comprises the deployment of new network automation technologies, bi-directional communication, smart metering, analysis and the extraction of value from massive amounts of data. This process enables the integration of further renewable-based generation, which contributes to the decarbonization of the economy but in turn creates new technical and market challenges due to its variable nature, and contributes also to the empowerment of consumers who may have a more proactive role through demand response mechanisms. The global aim is to develop more sustainable, reliable and efficient grids.

This Special Issue of *Energies* on "Optimization Models and Methods in Energy Systems" comprises nine papers, which have been recommended for publication after a thorough reviewing procedure. The papers address diversified problems, including the selection of offshore wind farm locations, a model for the analysis of regional energy system management and emission reduction potential, the management of building microgrid networks in islanded mode considering adjustable power and component outages, the portfolio optimization of power generation assets, a graph theoretic approach to optimal firefighting in oil terminals, a university building test case for occupancy-based building automation, the prospects of a meshed electrical distribution system featuring large-scale variable renewable power, a generation expansion planning model for an integrated energy system considering feasible operation region and the generation efficiency of combined heat and power, and the consideration of wind power generation in Brazil's long-term energy planning model.

Mytilinou et al. develop a framework to assist wind energy developers to select the optimal deployment site of a wind farm. The framework includes optimization techniques, decision-making methods and experts' input to support investment decisions. Techno-economic evaluation, life cycle costing and physical aspects for each location are considered along with experts' opinions to provide deeper insight into the decision-making process. A combination of the nondominated sorting multi-objective genetic algorithm NSGA-II and the multi-criteria technique for the order of preference by similarity to the ideal solution TOPSIS is used to recommend the optimal location. An application in the UK is reported to assist decision makers to make more informed and cost-effective decisions under uncertainty when investing in offshore wind energy.

Xie et al. propose a scenario-based multistage stochastic inexact robust programming for electric power generation planning and structure adjustment management under uncertainty. Uncertainties are expressed as interval values and probability distributions in the objective function and constraints. Power demand scenarios associated with electric power structure adjustment, imported electricity and emission reduction were designed to obtain multiple decision schemes for supporting the development of regional sustainable energy systems, in order to obtain a balanced development between conventional and new clear power generation technologies under uncertainty. This optimization framework was applied in Zibo City, Shandong Province, China.

Bui et al. develop an energy management scheme for islanded inter-connected building microgrid (BMG) networks, also considering an external energy supplier. Each BMG has a local combined heat and power (CHP) unit, energy storage, renewables and loads (electric and thermal). The external energy system comprises an external CHP unit, chillers, electric heat pumps and heat pile line, for thermal energy storage. The BMGs can trade energy with other BMGs in the network and with the external energy supplier. The adjustable power concept can reduce the operation cost of the network by increasing/decreasing the power of dispatchable units. The failure/recovery of components in the BMGs and the external system are also considered.

Glensk and Madlener present different fuzzy portfolio selection models, where the rate of returns as well as the investor's aspiration levels of portfolio return and risk are regarded as fuzzy variables. Portfolio risk is defined as a downside risk, leading to a semi-mean absolute deviation portfolio selection model. The models are applied to a selection of power generation mixes. The efficient portfolio results show that the fuzzy portfolio selection models with different definitions of membership functions, as well as the semi-mean absolute deviation model, perform better than the standard mean–variance approach. The consideration of membership functions for the description of investors' aspiration levels for the expected return and risk shows how the knowledge of experts, and investors' subjective opinions, can be better integrated into the decision-making process than with probabilistic approaches.

Khakzad presents a study to answer the question of fire brigades that need to optimally allocate their resources: "which burning units to suppress first and which exposed units to cool first?", until more resources become available from nearby industrial plants or residential communities. The aim is to contribute to more effective firefighting of major fires in fuel storage plants. A comparison between the outcomes of the graph theoretic approach and an approach based on influence diagrams shows the efficiency of the graph approach.

Swaminathan et al. propose a model-based approach to overcoming control fragmentation without disrupting the standard hierarchy of sub-systems of heating, ventilation and air-conditioning (HVAC) units in buildings, which must be accurately integrated and controlled by the building automation system to ensure the occupants' comfort. The set-point control is based on a predictive HVAC thermal model to optimize thermal comfort with reduced energy consumption. The standard low-level Proportional-Integral-Derivative (PID) controllers are auto-tuned based on simulations of the HVAC thermal model. Experimental and simulation validation is reported for university buildings in the Netherlands.

Cruz et al. present new operational strategies based on a meshed topology, aiming to increase the flexibility of power distribution system operators so that large amounts of intermittent renewable generation can be seamlessly accommodated. The distribution operational problem is formulated as a stochastic mixed-integer linear programming model. Numerical results reveal the multi-faceted benefits of operating distribution grids in a meshed manner. Such an operation scheme adds considerable flexibility to the system and leads to a more efficient utilization of variable renewable energy-distributed generation.

Ko and Kim propose a mixed-integer linear programming model for the minimization of the generation expansion cost of an integrated energy system, considering the feasible operation region and efficiency of a CHP resource to satisfy the varying demands of heat and electricity, which are interdependent and present different seasonal characteristics. Linearized constraints of a feasible operation region and the generation efficiency of the CHP resource are developed. A case study is presented, comparing the results of this model with those of a conventional optimization model that uses a constant heat-to-power ratio and generation efficiency of the CHP resource. Planning schedules and total generation efficiency profiles of the CHP resource for the compared optimization models are evaluated.

Maçaira et al. present an approach incorporating wind power generation in hydro-thermal dispatch using the analytical method of frequency and duration, with application to Brazil's Northeast region, covering the planning period from July 2017 to December 2021, using the Markov chain Monte Carlo method to simulate wind power scenarios. The results obtained are more conservative than the ones currently used by the National Electric System Operator, since the proposed approach forecasts 1.8% less wind generation, especially during peak periods, and 0.67% more thermal generation. This approach can reduce the chances of water reservoir depletion and an ineffective dispatch.

These papers offer a broad view of the relevant, diversified and challenging problems arising in energy systems, for which improved optimization models and methods should be developed and creatively applied.

I thank the authors and the reviewers for the care taken in preparing and assessing the papers, as well as the proficiency of the staff at MDPI. Moreover, I acknowledge the support of R&D projects ESGRIDS (POCI-01-0145- FEDER-016434) and MAnAGER (POCI-01-0145-FEDER-028040).

Carlos Henggeler Antunes
Special Issue Editor

![energies logo] **energies**

MDPI

Article

A Framework for the Selection of Optimum Offshore Wind Farm Locations for Deployment

Varvara Mytilinou [1,*], Estivaliz Lozano-Minguez [2] and Athanasios Kolios [3]

[1] Renewable Energy Marine Structures Centre for Doctoral Training, Cranfield University, Cranfield, Bedfordshire MK43 0AL, UK
[2] Department of Mechanical Engineering and Materials—CIIM, Universitat Politècnica de València, Camino de Vera s/n, 46022 Valencia, Spain; eslomin@upv.es
[3] Department of Naval Architecture, Ocean & Marine Engineering, University of Strathclyde, HD2.35, Henry Dyer Building, 100 Montrose Street, Glasgow G4 0LZ, UK; athanasios.kolios@strath.ac.uk
* Correspondence: v.mytilinou@cranfield.ac.uk; Tel.: +44-1234-75-4631

Received: 23 April 2018; Accepted: 9 July 2018; Published: 16 July 2018

Abstract: This research develops a framework to assist wind energy developers to select the optimum deployment site of a wind farm by considering the Round 3 available zones in the UK. The framework includes optimization techniques, decision-making methods and experts' input in order to support investment decisions. Further, techno-economic evaluation, life cycle costing (LCC) and physical aspects for each location are considered along with experts' opinions to provide deeper insight into the decision-making process. A process on the criteria selection is also presented and seven conflicting criteria are being considered for implementation in the technique for the order of preference by similarity to the ideal solution (TOPSIS) method in order to suggest the optimum location that was produced by the nondominated sorting genetic algorithm (NSGAII). For the given inputs, Seagreen Alpha, near the Isle of May, was found to be the most probable solution, followed by Moray Firth Eastern Development Area 1, near Wick, which demonstrates by example the effectiveness of the newly introduced framework that is also transferable and generic. The outcomes are expected to help stakeholders and decision makers to make better informed and cost-effective decisions under uncertainty when investing in offshore wind energy in the UK.

Keywords: multi-objective optimization; nondominated sorting genetic algorithm (NSGA); multi-criteria decision making (MCDM); technique for the order of preference by similarity to the ideal solution (TOPSIS); life cycle cost

---

1. Introduction

The future of wind energy seems to keep growing as 18 GW are expected to be deployed by 2020 in the UK, with potential for more ambitious targets after 2020. Thus, there is a substantial need to reduce the cost of energy by identifying relevant cost reduction strategies in order to achieve these goals. The future of the UK's industry size strongly depends on these goals [1]. Significant price increases in the overall cost of turbines, operations and maintenance have a direct impact on large-scale wind projects, hence the wind energy industry is determined to lower the costs of producing energy in all phases of the wind project from predevelopment to operations. Following the UK technology roadmap, the offshore wind costs should be reduced to £100/MWh by 2020 [2]. According to [1] the costs were stabilized at £140 per MWh in 2011. The UK's Offshore Wind Programme Board (OWPB) stated that the offshore wind costs dropped below £100/MWh when 2015–2016 projects achieved a levelized cost of energy (LCOE) of £97 compared to £142 per MWh in 2010–2011, according to the Cost Reduction Monitoring Framework report in 2016 [3]. Recently, in 2017, Ørsted (formerly DONG

Energy) guaranteed a £57.5/MWh building the world's largest offshore wind farm in Hornsea 2, according to [4].

Developers and operators of offshore wind energy projects face many risks and complex decisions regarding service life cost reduction. In many cases, the manufacturers produce large volumes of parts in order to deal with the issue via economies of scale. Also, project consents can be time-consuming and difficult to obtain, however, all offshore wind farms were successfully completed regarding investment and profit [1]. Ensuring a long-term and profitable investment plan can be challenging, with both pre-consent and post-consent delays introducing considerable risks [2,5]. To this end, appropriate planning studies should be conducted at the early development stages of the project in order to minimize the investment risk. A breakdown of the key costs in an offshore wind farm can be found in [6] while studying existing projects, the location of a wind farm and the type of support structure have a great impact on the overall costs [7–9].

The aim of this paper is to develop a wind farm deployment framework, as illustrated in Figure 1, for supporting investment decisions at the initial stages of the development of Round 3 offshore wind farms in the UK by combining multi-objective optimization (MOO), life cycle cost (LCC) analysis and multicriteria decision making (MCDM). The contribution to knowledge is in developing and applying this novel and transferable framework that combines an economic analysis model by using LCC and geospatial analysis, MOO by using nondominated sorting genetic algorithm (NSGA II), survey data from real-world experts and finally MCDM by using a deterministic version and a stochastic expansion of the technique for the order of preference by similarity to the ideal solution (TOPSIS). Also, a criteria selection framework for the implementation of MCDM methods has been devised. The outcomes are expected to provide a deeper insight into the wind energy sector for future investments.

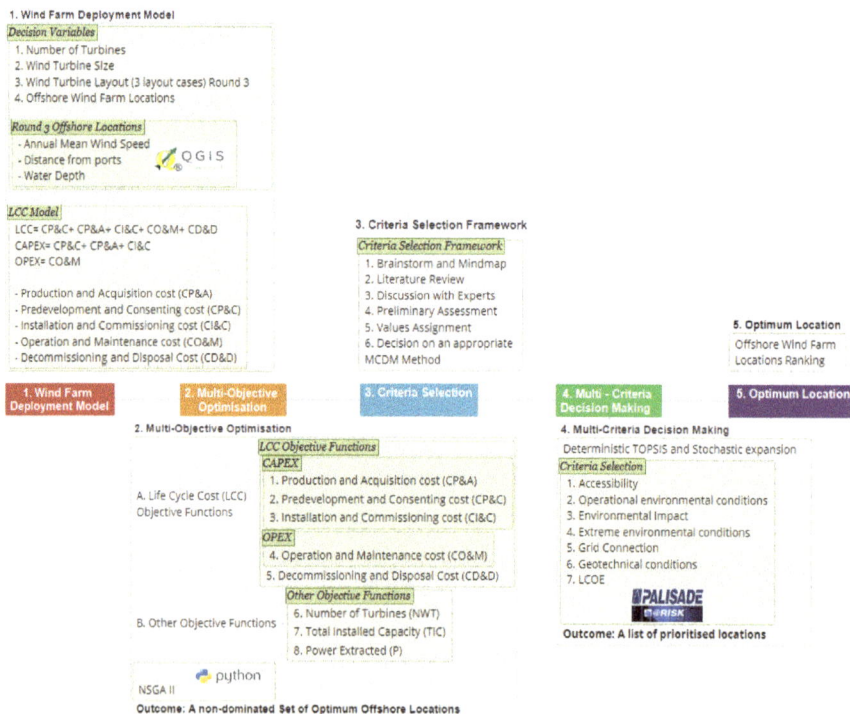

Figure 1. Main framework.

The structure of the remaining sections of this paper starts with a literature review on related studies for LCC analysis, turbine layout optimization, MCDM, and wind farm location selection in the offshore wind energy sector. Next, the development of the proposed framework is documented. The nondominated results for all zones will be analyzed and discussed followed by the prioritization process from TOPSIS. Conclusions and future work are documented at the end of the paper.

2. Literature Review

The Crown Estate has the rights of the seabed leasing up to 12 nautical miles from the UK shore and the right to exploit the seabed for renewable energy production up to 200 miles across its international waters. In recent years, the Crown Estate has run three rounds of wind farm development sites and their extensions. When the Crown Estate released the new Round 3 offshore wind site leases, they provided nine large zones of up to 32 GW power capacity [10]. The new leases encourage larger scale investments and consequently bigger wind turbines and include locations further away from the shore and in deeper waters [2,5,11–13].

Currently, all Round 3 zones have been suggested and published according to reports by the Department for Energy and Climate Change (DECC) and other stakeholders after the outcome of a strategic environmental assessment [14]. It should be noted that new offshore and onshore electricity transmission networks are needed in order to cover Round 3 connections up to 25 GW [14]. The Round 3 zones are the following; Moray Firth, Firth of Forth, Dogger Bank, Hornsea, East Anglia (Norfolk Bank), Rampion (Hastings), Navitus Bay (West Isle of Wight), Atlantic Array (Bristol Channel), and Irish Sea (Celtic Array). Every zone consists of various sites and extensions. Here, the five first zones in the North Sea are investigated in order to demonstrate the proof of the developed framework's applicability. Each location faces similar challenges such as deep waters or long distances from the shore, etc. as shown in Figure 2.

Only a few location-selection-focused studies can be found, and usually, the findings and the formulation of the problems follow a different direction that this present study. For instance [15], uses goal programming in order to obtain the optimum offshore location for a wind farm installation. The study involves Round 3 locations in the UK and discusses its flexibility to combine decision-making. The work integrates the energy production, costs and multicriteria nature of the problem while considering environmental, social, technical and economic aspects.

For instance, the following literature presents cases in renewable energy where optimization has been successfully applied by utilizing different algorithms. An approach that links a multi-objective genetic algorithm to the design of a floating wind turbine was presented in [16]. By varying nine design variables related to the structural characteristics of the support structure, multiple concepts of support structures were modelled and linked to the optimizer. In [17], the authors provide a case study for the optimization of the electricity generation mix in the UK by using hybrid MCDM and linear programming and suggest a methodology to deal with the uncertainty that is introduced in the problem by the bias in experts' opinions and other related factors. In [18], a structural optimization model for the support structures of offshore wind turbines was implemented by using a parametric Finite Element Analysis (FEA) analysis coupled with a genetic algorithm in order to minimize the mass of the structure considering multicriteria constraints.

LCC analysis evaluates costs, enabling suggestions in cost reductions throughout a project's service life. The outcome of the analysis provides pertinent information in investments and can influence decisions from the initial stages of a new project [19]. In [20], a parametric whole life cost framework for an offshore wind farm and a cost breakdown structure was presented and analyzed, where the project is divided into five different stages; the predevelopment and consenting ($C_{P\&C}$), production and acquisition ($C_{P\&A}$), installation and commissioning ($C_{I\&C}$), operation and maintenance ($C_{O\&M}$), and decommissioning and disposal ($C_{D\&D}$) stage. The advantages and disadvantages of the transition to offshore wind and an LCC model of an offshore wind development were proposed in [21]. However, the study mainly focused on a simplified model and especially the operation and maintenance stage of the LCC analysis, and it was suggested that there could be a further full-scale LCC framework in the future. In [22], a detailed failure mode identification throughout the service life of offshore wind turbines was performed and a review of the three most relevant end-of-life scenarios were presented in order to contribute to increase the return on investment and decrease the levelized cost of electricity. However, there are limited studies that integrate a high fidelity of life cycle cost (LCC) analysis into a multi-objective optimization (MOO) algorithm. LCC analysis gains more ground over the years because of the increased uncertainty of wind energy projects throughout their service life, including the cost of finance, the real cost of Operational Expenditure (OPEX) and the potential of service life extension.

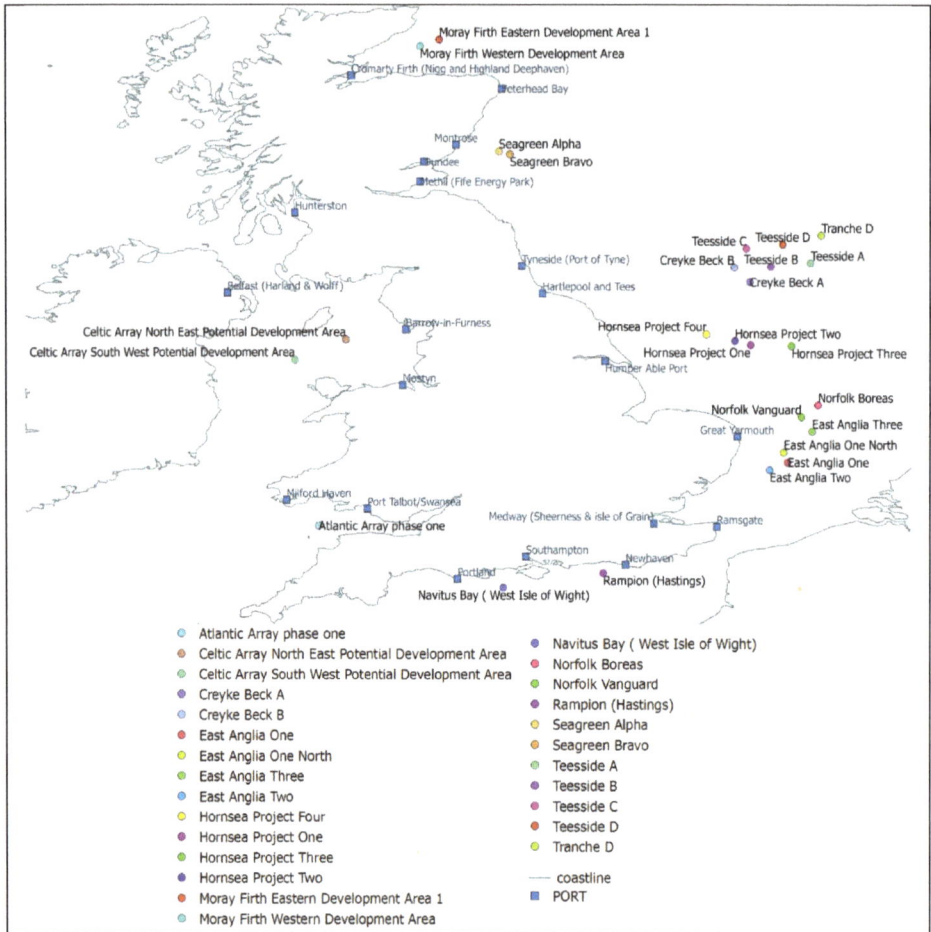

Figure 2. Round 3 offshore locations around the UK by using open source licensed geographic information system QGIS.

MCDM is beneficial for policy-making through evaluation and prioritization of available technological options because of their ability to combine both technical and non-technical alternatives as well as quantitative and qualitative attributes in the decision-making process. A number of MCDM methods are applicable to energy-related projects, however, TOPSIS was selected because of the wide applicability of the method as can be found in literature and the connection of the method to numerous energy-related studies such as [23–25]. It is common to combine stochastic and fuzzy processes in order to deal with uncertain environments. In [23], Lozano-Minguez employed a methodology on the selection of the best support structure among three design options of an offshore wind turbine, considering a set of qualitative and quantitative criteria. A similar study was reported by Kolios in [26], extending TOPSIS to consider stochasticity of inputs.

Methods and techniques to cope with a high number of criteria and high dimensionality of decision-making problems are available in the literature. The multiple criteria hierarchy process (MCHP) [27–29] has been employed in order to deal with multiple criteria in decision-making processes. MCHP is usually employed in combination with outranking MCDM methods. Further applications can be found in [30,31].

In general, classifying criteria as either qualitative or quantitative is related to their nature and fidelity of the analysis. The employed decision-making methods can be based on priority, outranking, distance or combination of the three [32]. In [23], a decision-making study was conducted in three fixed wind turbine support structure types considering both quantitative and qualitative criteria while using TOPSIS. A decision-making study on floating support structures by combining both quantitative and qualitative criteria was presented in [33].

The approach proposed here for the stochastic expansion of deterministic methods was based in [26] that has reported the expansion of different deterministic methods, under the consideration that input variables are modelled as statistical distributions (derived by fitting data collected for each value in the decision matrix and weight vector), as shown in Figure 3. By using Monte Carlo simulations, numerous iterations quantify results and identify the number of cases where the optimum solution will prevail, i.e., there is a P_i probability that option X_i will rank first.

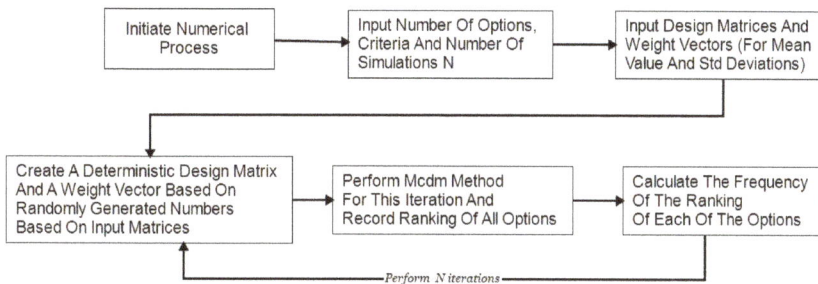

Figure 3. Stochastic expansion algorithm of deterministic Multi-Criteria Decision Making (MCDM) methods.

In [26], during deterministic TOPSIS, the weights for each criterion were considered fixed, but under stochastic modelling, statistical distributions were employed to best fit the acquired data of the experts' opinions. Perera [34] has presented a study that combines MCDM and multi-objective optimization in the designing process of hybrid energy systems (HESs), using the fuzzy TOPSIS extension along with level diagrams. In [35], MCDM under uncertainty is discussed in an application where the alternatives' weights are partially known. An extended and modified stochastic TOPSIS approach was implemented using interval estimations.

In [26], the authors extend the previous MCDM study on the decision-making of an offshore wind turbine support structure among different fixed and floating types. The decision matrix includes stochastic inputs (by using data from experts) in order to minimize the uncertainties in the study. In the same study, an iterative process has been included, and the TOPSIS method was implemented. In [36], a study suggests a methodology for classification and evaluation of 11 available offshore wind turbine support structure types while considering 13 criteria by using TOPSIS as the decision-making method.

In [24], an expansion of MCDM methods to account for stochastic input variables was conducted, where a comparative study was carried out by utilizing widely applied MCDM methods. The method was applied to a reference problem in order to select the best wind turbine support structure type for a given deployment location. Data from industry experts and six MCDM methods were considered, so as to determine the best alternative among available options, assessed against selected criteria in order to provide a level of confidence to each option.

An electricity generation systems allocation optimization model is suggested in [37] for the case of a disaster relief camp in order to minimize the total project cost and maximize the share of systems that were assessed through a decision-making process and were prioritized accordingly. Bi-objective integer linear programming and a decision-making method (VIKOR) were employed and the overall model was applied to a hypothetical map.

A study performed in [38] uses a TOPSIS model by incorporating technical, environmental and social criteria and finally combines the evaluation scores to develop a MOGLP (multi-objective grey linear programming) problem in order to assess the decision-making of power production technologies. The outcome of this work was the optimal mix of electricity generated by each option in the UK energy market. In [39], a methodology for an investment risk evaluation and optimization is suggested in order to mitigate the risks and achieve sustainability for wind energy projects in China. In this study, Monte Carlo analysis and a multi-objective programming model are used so as to increase the confidence in the planning of investment research and the sustainability of renewables in China.

In this study, NSGA II is employed because it is suitable for MOO problems with many objectives and was further analyzed in previous studies in offshore wind energy applications in [40], where a methodology was proposed to support the decision-making process at these first stages of a wind farm investment considering available Round 3 zones in the UK. Three state-of-the-art algorithms were applied and compared to a real-world case of the wind energy sector. Optimum locations were suggested for a wind farm by considering only round 3 zones around the UK. The problem comprised of techno-economic Life Cycle Cost related factors, which were modelled by using the physical aspects of each wind farm location (i.e., the wind speed, distance from the ports and water depth), the wind turbine size and the number of turbines.

3. Framework

3.1. Wind Farm Deployment Model

The wind farm deployment model implemented in this study couples the LCC analysis with a geospatial analysis as described below. The LCC analysis of a project involves all project stages described in Figure 4. In [20,41], a whole LCC formulation is provided, and this study integrates these phases into the MOO problem. Assumptions and related data in the modelling of the problem were gathered from the following references [20,41–46] based on which the present model was developed. The LCC model described in [20] is used as a guideline in this study, and along with the site characteristics and the problem's formulation, the optimization problem is formed. The type of foundation that was considered in the LCC model is the jacket structure as it constitutes a configuration that can be utilized in a range of water depths allowing for the optimization process to be automated. The total LCC is calculated as follows:

$$LCC = C_{P\&C} + C_{P\&A} + C_{I\&C} + C_{O\&M} + C_{D\&D} \qquad (1)$$

where

LCC: Life Cycle Cost
$C_{I\&C}$: Installation and Commissioning cost
$C_{P\&C}$: Predevelopment and Consenting cost
$C_{O\&M}$: Operation and Maintenance cost

$C_{P\&A}$: Production and Acquisition cost

$C_{D\&D}$: Decommissioning and Disposal Cost

$$CAPEX = C_{P\&C} + C_{P\&A} + C_{I\&C} \tag{2}$$

$$OPEX = C_{O\&M} \tag{3}$$

CAPEX: Capital expenditure

OPEX: Operational expenditure

The power extracted is calculated for each site and each wind turbine respectively as:

$$P = \frac{1}{2}ACp\rho u^3 \tag{4}$$

where

A : Turbine rotor area

ρ: Air density

C_p: Power coefficient

u : Mean annual wind speed of each specific site

The total installed capacity (TIC) of the wind farm dependents on the number of turbines and the rated power of each of them, and is calculated for every solution:

$$TIC = P_R \times NWT \tag{5}$$

where

P_R: Rated power

NWT: Number of turbines

For each offshore location, a special profile was created including the coordinates, distance from designated construction ports, annual wind speed and average site water depth, as listed in Table 1, where data was acquired from [45]. Among various data, Table 1 shows the locations that each of these zones contains.

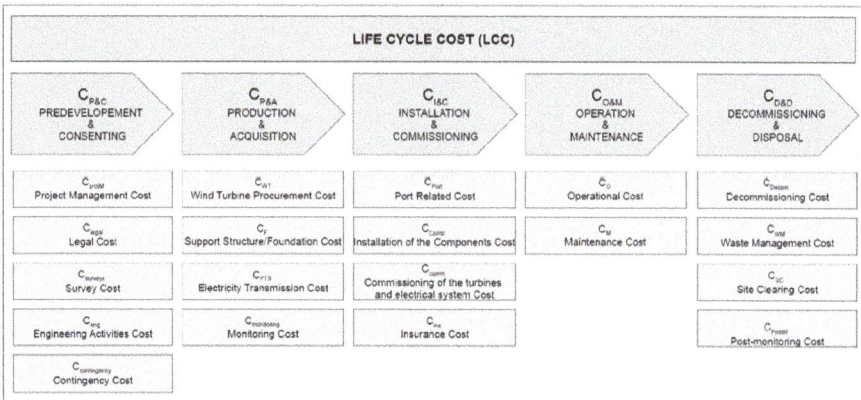

Figure 4. Life cycle cost (LCC) breakdown [20].

Table 1. Round 3 zones and sites, and specific data acquired from [45].

Site Index	Zone	Wind Farm Site Name	Centre Latitude	Centre Longitude	Port	Distance from the Port (km)	Annual Wind Speed (m/s) (at 100 m)	Average Water Depth (m)
0	Moray Firth	Moray Firth Western Development Area	58.097	−3.007	Port of Cromarty	123.691	8.82	44
1	Moray Firth	Moray Firth Eastern Development Area 1	58.188	−2.720	Port of Cromarty	157.134	9.43	44.5
2	Firth of Forth	Seagreen Alpha	56.611	−1.821	Montrose	72.598	9.92	50
3	Firth of Forth	Seagreen Bravo	56.572	−1.658	Montrose	91.193	10.09	50
4	Dogger Bank	Creyke Beck A	54.769	1.908	Hartlepool and Tess	343.275	10.01	21.5
5	Dogger Bank	Creyke Beck B	54.977	1.679	Hartlepool and Tess	319.949	10.04	26.5
6	Dogger Bank	Teesside A	55.039	2.822	Hartlepool and Tess	447.124	10.05	25.5
7	Dogger Bank	Teesside B	54.989	2.228	Hartlepool and Tess	380.788	10.04	25.5
8	Hornsea	Hornsea Project One	53.883	1.921	Grimsby	242.328	9.69	30.5
9	Hornsea	Hornsea Project Two	53.940	1.687	Grimsby	217.270	9.73	31.5
10	Hornsea	Hornsea Project Three	53.873	2.537	Grimsby	310.521	9.74	49.5
11	Hornsea	Hornsea Project Four	54.038	1.271	Grimsby	173.928	9.71	44.5
12	East Anglia (Norfolk Bank)	East Anglia One	52.234	2.478	Great Yarmouth	92.729	9.5	35.5
13	East Anglia (Norfolk Bank)	East Anglia One North	52.374	2.421	Great Yarmouth	81.104	9.73	45.5
14	East Anglia (Norfolk Bank)	East Anglia Two	52.128	2.209	Great Yarmouth	74.559	9.46	50
15	East Anglia (Norfolk Bank)	East Anglia Three	52.664	2.846	Great Yarmouth	124.969	9.56	36
16	East Anglia (Norfolk Bank)	Norfolk Boreas	53.040	2.934	Great Yarmouth	143.464	9.53	31.5
17	East Anglia (Norfolk Bank)	Norfolk Vanguard	52.868	2.688	Great Yarmouth	111.449	9.56	32

For the distances from the ports calculation an open source licensed geographic information system (GIS) called QGIS was used, which is a part of the Open Source Geospatial Foundation (OSGeo) [47]. A list of ports was acquired from [48–50]. The QGIS maps of the offshore sites were acquired from the official Crown Estate website [51] for QGIS and AutoCAD. The list contains designated, appropriate and sufficient construction ports that are suitable for the installation, manufacturing and maintenance for wind farms. New ports are to be built specifically to accommodate needs of the offshore wind industry; however, this study takes into account a selection of currently available ports around the UK. The distances were calculated assuming that the nearest port to the individual wind farm is connected in a straight line. QGIS was also employed to measure and model aspects of the LCC related to the geography and operations. The estimated metrics were integrated into the configuration settings of the whole LCC.

Three layout configurations are considered. The lower and upper limits of a theoretical array layout from [52] will be employed along with an extreme case. More specifically, in the lower limit case (layout 1), the horizontal and vertical distance between turbines is 3 and 5 times the rotor diameter, respectively. The turbine specifications used for the LCC model are listed in Table 2. In the upper limit case (layout 2), 5 and 9 times the rotor diameter were considered horizontally and vertically. In the extreme case (layout 3), the horizontal and vertical distance between turbines is 10 and 18 times the rotor diameter. All different configurations are illustrated in Figure 5. The present work focuses on the optimization of offshore wind farm locations considering the maximum wind turbine number that can fit in the selected Round 3 locations according to three different layout configuration placements. The wind farm is oriented according to the most optimal wind direction. Different layouts provide a different maximum wind turbine number that can guide the optimization process to more detailed calculations. The maximum number of wind turbines is determined by considering types of reference turbines of 6, 7, 8 and 10 MW and by following three layout cases, as listed below in Figure 5, where D is the diameter of each turbine.

Table 2. Turbine specifications.

Turbine Type Index	Rated Power (MW)	Rotor Diameter (m)	Hub Height (m)	Total Weight (t)
0	10	190	125	1580
1	8	164	123	965
2	7	154	120	955
3	6	140	100	656

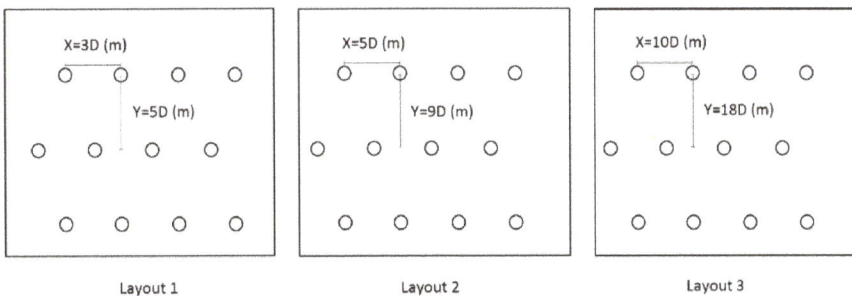

Figure 5. Demonstrating different layouts, where D corresponds to the diameter of the turbine.

For the estimation of cabling length, which is required to calculate parts of the LCC related to the spatial distribution of the wind turbines in the wind farm, the minimum spanning tree algorithm is used. The location of the turbines is treated as a vertex of a graph, and the cabling represents the edge that connects the vertices. Given a set of vertices, which are separated by each other by the different layout indices, from Figure 5, the minimum spanning tree connects all these vertices without creating any cycles, thus yielding minimum possible total edge length. This represents the minimum cabling length of the particular layout.

The way the length of the cables was calculated provides an approximation of the actual length. In the presence of relevant actual data, the calculations of both the layouts and the LCC would provide more realistic values. For instance, the cable length would be expected to be larger because of the water depth and the burial of the cables for each turbine. For each cable, both ends will have to come from the seabed to the platform, so at least twice the water depth should be added to each cable and finally allow for some contingency length for installation.

The wind rose diagrams provided the prevailing wind direction, which sets the layout orientation. The wind speeds, the wind rose graphs, and the coordinates of each location were obtained by FUGRO (Leidschendam, The Netherlands) and 4COffshore (Lowestoft Suffolk, UK) [45,53]. All wind farm sites were discovered to have dominant southwestern winds followed by western winds. For that reason, the orientation of the layouts is assumed to be southwestern (as the winds are assumed to blow predominantly from that direction). The wind rose graphs for each offshore site are determined by data acquired from [53] and the grid points they created around the UK. The nearest grid point to the offshore site is used.

An important factor to be considered is also the atmospheric stability. Although the different layouts considered in this study may be affected by the atmospheric stability states, as it impacts the layout's wake recovery pattern, it was not considered in the framework. Also, the power curves and their multiplicity in turbine type were not considered in this study because the aim is to devise and demonstrate a generic and transferable methodology. It is suggested that both elements could be further investigated in future studies to evaluate their effect in the derivation of the optimum solution.

In Figure 6, the example of Moray Firth zone (which includes Moray Firth Western Development Area and Moray Firth Eastern Development Area 1) shows the positioning of the turbines depending on the layout 1, 2 and 3 and the turbine size.

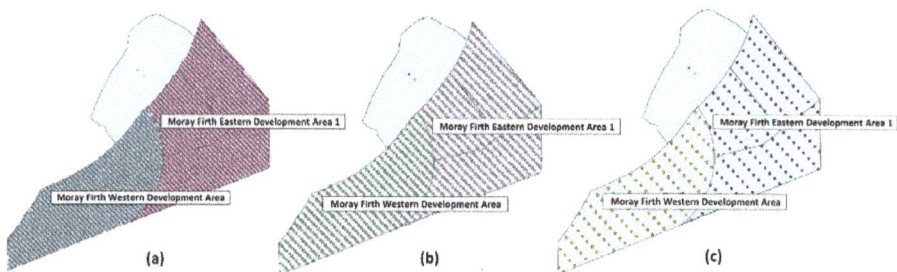

Figure 6. Moray Firth zone. A maximum number of wind turbines placed according to layout 1, layout 2 and layout 3 for the case of 10 MW turbine. In (**a**) Moray Firth, 10 MW turbines positioned in layout 1; (**b**) Moray Firth, 10 MW turbines positioned in layout 2; (**c**) Moray Firth, 10 MW turbines positioned in layout 3.

3.2. Multi-Objective Optimization

The optimization problem includes eight objectives; five LCC-related objectives, based on [20], which are the cost-related objectives to be minimized. The three additional objectives are the number of turbines (NWT), the power that is extracted (P) from each offshore site and the total installed capacity (TIC), which are to be minimized, maximized, and maximized, respectively.

More specifically, the LCC includes the predevelopment and consenting, production and acquisition, installation and commissioning, operation and maintenance and finally decommissioning and disposal costs. The power extracted is calculated by the specific mean annual wind speed of each location along with the characteristics of each wind turbine both of which are considered inputs.

The optimization problem formulates as follows:

$$\text{Minimize} \quad C_{P\&C}, C_{P\&A}, C_{I\&C}, C_{O\&M}, C_{D\&D}, NWT, (-P), (-TIC) \tag{6}$$

Subject to $0 \le$ site index ≤ 20,

$\quad\quad 0 \le$ turbine type index ≤ 3

$\quad\quad 1 \le$ layout index ≤ 3

$\quad\quad 50 \le$ Number of turbines $\le$ maximum number per site

$\quad\quad TIC \le$ Maximum capacity of Round 3 sites based on the Crown Estate

Although the maximum number of turbines has been estimated by using QGIS, the maximum capacity allowed per region was also considered, as specified by the Crown Estate, as listed in Table 3. These were selected because of the possibility that the constraints might overlap in an extreme case scenario. Therefore, both constraints were added to the problem in order to secure all cases.

Table 3. Maximum capacity of Round 3 wind farms, specified by the Crown Estate [1].

Zone	Capacity (MW)
1. Moray Firth	1500
2. Firth of Forth	3465
3. Dogger Bank	9000
4. Hornsea	4000
5. East Anglia	7200
6. Rampion	665
7. Navitus Bay	1200
8. Bristol Channel	1500
9. Celtic Array	4185
TOTAL CAPACITY	32,715

The optimization part of the framework has been implemented in Python 3, employing library 'platypus' in Python [54].

3.3. Criteria Selection Process

For the MCDA, the criteria selection process follows the process illustrated in Figure 7:

1. The first step is to create a mind map of the problem and the different aspects involved. Then, via brainstorming, criteria that can potentially impact on the alternatives of the problem are listed.
2. The second step is to perform an extensive literature review on the topic. It is vital that the literature review is conducted in order to discover related studies and also confirm or reject ideas that were found in the first step. During this process, it is possible to discover gaps that will help to define the study more precisely and also discover criteria that were never considered before.
3. Step three is about discussing ideas with subject matter experts and communicating to them the aims and ideas of the project in order to obtain useful insights into the initial stages of the criteria selection. Their expertise can confirm, discard or suggest new criteria according to their opinion. Experts can also provide helpful data and confirm the value of the study.
4. In step four, the strengths and weaknesses of the work and criteria should be identified, followed by a preliminary assessment. The selected criteria should be clear and precise, and no overlaps should be present (avoiding similar terms or definitions that can potentially include other criteria). Each criterion should characterize and affect the alternatives in a different and unique way. None of the criteria should conflict with each other. The criteria should now have a detailed description. Their description and explanation should be unique to avoid confusion especially if the criteria are sent to experts in the form of a survey.

5. Step five describes how to proceed with the study. Assigning values to the criteria can be done either by calculating the values directly or by extracting them from the experts via a questionnaire. In the latter case, additional data or opinions could be considered. Via a survey, experts could either assign values or rate the criteria according to their knowledge and experience. Here, it is important to note that for a different set of criteria, different approaches can be followed. For example, in the case of criteria that need numerical values (and probably require calculations) that no expert can provide on the spot, receiving replies is challenging. The experts should provide their expertise in an easy and fast process. The definition of the criteria has to be very clear before scoring, normally at a scale of 1 to 5 or otherwise. The calculations could lead to assigned values for every criterion, but the experts could provide further insight regarding the importance of those criteria and how much they affect the alternatives. In this case, the experts provide the weights of the criteria, which is very useful in order to achieve higher credibility of the problem. In some cases, it would be very useful to include validation questions in the survey. It would also be useful to include questions in order to increase the validity of the problem, for example, to ask for further criteria that were not considered in the study. Another example would be to include a question about the perceived expertise of the experts that will answer the questionnaire. Hence, their answers will be weighted and further credible.

6. Step six is related to selecting a method for decision-making. In general, it is important to decide quite early which method of the multicriteria analysis will be used. This is important because different methods require different criteria and problem set up. In the case of hierarchy problems and pairwise comparisons, the problem has to be set up differently, and the values need to be set for every pair comparison. The important question here is how the outcomes are derived. Having a picture of the total process and aims, objectives and results early enough can help to speed up the process.

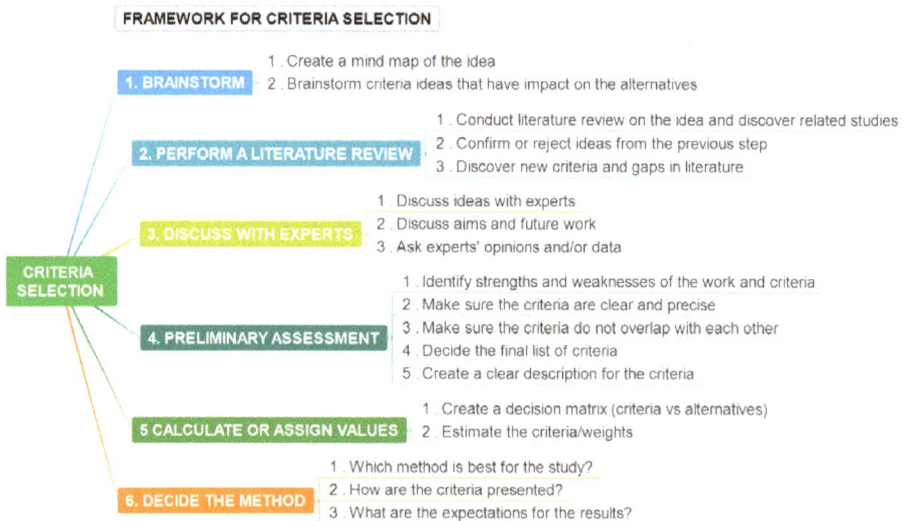

Figure 7. Criteria selection framework.

3.4. Multicriteria Decision Making

Following the process of MOO and criteria selection, two versions of the MCDM method were implemented (i.e., deterministic and stochastic TOPSIS) and were linked to the results of the previous outcomes, as shown in Figure 1. A set of qualitative and quantitative criteria is combined in order to investigate the diversity and outcomes obtained from different sets of inputs in the decision-making process. Stochastic inputs are selected and imported in TOPSIS. All data were collected from industry experts, so as to prioritize the alternatives and assess them against seven selected conflicting criteria. The outcome of the method is expected to assist stakeholders and decision makers to support decisions and deal with uncertainty, where many criteria are involved.

TOPSIS is depicted in Figure 8, initially proposed by Hwang et al. [55], and the idea behind it lies in the optimal alternative being as close in the distance as possible from an ideal solution and at the same time as far away as possible from a corresponding negative ideal solution. Both solutions are hypothetical and are derived from the method. The concept of closeness was later established and led to the actual growth of the TOPSIS theory [56,57].

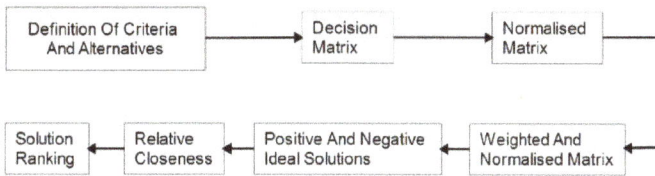

Figure 8. TOPSIS methodology.

After defining n criteria and m alternatives, the normalized decision matrix is established. The normalized value r_{ij} is calculated from the equations below, where f_{ij} is the i-th criterion value for alternative A_j ($j = 1, \ldots, m$ and $i = 1, \ldots, n$).

$$r_{ij} = \frac{f_{ij}}{\sqrt{\sum_{j=1}^{m} f_{ij}^2}} \tag{7}$$

The normalized weighted values v_{ij} in the decision matrix are calculated as follows:

$$v_{ij} = w_i r_{ij} \tag{8}$$

The positive ideal A^+ and negative ideal solution A^- are derived as shown below, where I' and I'' are related to the benefit and cost criteria (positive and negative variables).

$$A^+ = \{v_1^+, \ldots, v_n^+\} = \{(MAX_j v_{ij} | i \in I'), (MIN_j v_{ij} | i \in I'')\} \tag{9}$$

$$A^- = \{v_1^-, \ldots, v_n^-\} = \{(MIN_j v_{ij} | i \in I'), (MAX_j v_{ij} | i \in I'')\} \tag{10}$$

From the n-dimensional Euclidean distance, D_j^+ is calculated below as the separation of every alternative from the ideal solution. The separation from the negative ideal solution follows:

$$D_j^+ = \sqrt{\sum_{i=1}^{n} (v_{ij} - v_i^+)^2} \tag{11}$$

$$D_j^- = \sqrt{\sum_{i=1}^{n} (v_{ij} - v_i^-)^2} \tag{12}$$

The relative closeness to the ideal solution of each alternative is calculated from:

$$C_j = \frac{D_j^-}{\left(D_j^+ + D_j^-\right)} \tag{13}$$

After sorting the C_j values, the maximum value corresponds to the best solution to the problem.

A survey that considers all seven criteria was created and disseminated to industry experts, so as to obtain the weights for the following MCDM study. In this case, experts provided their opinions based on the importance of each criterion in the wind farm location selection process. In total, 13 experts (i.e., academics, industrial experts and university partners) with relative expertise responded and rated the criteria according to their importance. The total number of 13 experts is considered sufficient for this work because the overall number of offshore wind experts is very limited and their engagement is challenging. The input data from the 13 experts were acquired through an online survey platform where the perceived level of expertise was also provided. The assessments varied between 2 and 5 (with 1 being a non-expert and 5 being an expert) with a mean value of 3.8 and a standard deviation of 0.89.

The implementation of the stochastic version of TOPSIS was modelled through Palisade's software @Risk 7.5. Specifically, for the stochastic implementation, the Monte Carlo simulations of @Risk were combined with the survey data, providing the best distribution fit for each value to be used as inputs in the decision matrix of TOPSIS. By separately conducting a sensitivity analysis among 100, 1000, 10,000 and 100,000 iterations, 10,000 iterations for a simulation were found to deliver satisfactory results on acceptable computational effort requirements. Next, the stochastic approach is compared to the deterministic one and finally, the outcomes are presented in the next section.

All criteria and the final decision-making matrices were scaled and normalized, respectively in different phases of the process, while the seven criteria used in this study include both qualitative and quantitative inputs. Combining these two types can help decision makers to define their problems in a more reliable method. Next, both deterministic and stochastic approaches will be conducted and compared. The criteria are listed in Table 4.

Table 4. List of criteria.

Criteria	ID
1. Accessibility	C1
2. Operational environmental conditions	C2
3. Environmental Impact	C3
4. Extreme environmental conditions	C4
5. Grid Connection	C5
6. Geotechnical conditions	C6
7. LCOE	C7

The criteria were selected based on literature and a brainstorming session with academic and industrial experts. In the session, common criteria were consolidated in order to avoid double counting and finally concluded to the ones used in the study. The criteria were selected such as to have both a manageable number and to cover all aspects but at the same time not make the data collection questionnaire too onerous.

More specifically, the criteria are defined and analyzed below:

1. Accessibility: This criterion considers the accessibility of each wind farm by considering the distance from the ports and the number of nearby wind farms. The distances were acquired from the 4COffshore database [58]. The number of nearby wind farms was acquired from the interactive map of 4COffshore [45]. In order to select the number of nearby farms, only the farms that already produce energy and are located between the ports and the wind farm in question were considered. The nearby wind farms and the distance from the ports were assessed from 1 to 9 (1 being not close to any wind farms and 9 being close to many wind farms) and 9 to 1 (9 being very close to the ports and 1 being extremely far from the shore) respectively for each offshore site. The weighted values (equally weighted by 50–50) then were summed. This criterion is qualitative, and it varies from 1 to 9 (1 being not at all accessible to 9 extremely accessible).

This criterion is also considered positive in the MCDM process. Both in the deterministic and stochastic processes, the values used are the same.

2. Operational environmental conditions: This criterion considers the aerodynamic loads in the deployment location. More specifically, the wind speed (m/s) in specific points (close to each offshore sites) according to [53]. The criterion is quantitative and also positive. In the stochastic and deterministic approach, the fitted wind distributions and the mean values were used, respectively.

3. Environmental impact: This criterion considers the structures' greenhouse gas emissions during the construction and installation phase. The amount of CO_2 equivalent (CO_2e) emissions per kg of steel was estimated relative to the water depth (maximum and minimum water depth were measured in each location) and the distance from the ports. The support structure was assumed to be the jacket structure. This criterion was calculated according to an empirical formula in [23], and the water depth and distance from the ports were both considered in these calculations. Finally, an index of the square of CO_2 equivalent (CO_2e^2) was considered from the two cases as a value for each offshore site. This criterion is negative. The criterion is also quantitative, and for the stochastic approach, a triangle distribution was considered. In the deterministic approach, the mean value was used.

4. Extreme environmental conditions: This criterion considers the durability of the structure due to extreme aerodynamic environmental loads. Data were extracted from [53]. The wind distributions that represent the probabilities above the cut off wind speed (i.e., approximately 25 m/s) were considered. This criterion is quantitative and negative. For the stochastic approach, a triangle distribution was considered. In the deterministic approach, the mean values were used.

5. Grid connection: This criterion considers the possible grid connection options of a new offshore wind farm (connection costs to existing or new grid points). The inputs of this criterion consider the cost (£million) of connecting to nearby substations where other Rounds already operate, extending existing ones or building new ones. In the national grid report that was created for the Crown Estate in [14], the costs were calculated by considering more than one cases per Round 3 location. In this study, the maximum and the minimum costs were considered, and a uniform distribution was used as a stochastic input. In the deterministic approach, the mean value is used. The criterion is quantitative and represented by the above cost values, and it is considered negative.

6. Geotechnical conditions: This criterion represents the compatibility of the soil of each of the offshore locations for a jacket structure installation. Experts provided their input and rated the offshore locations according to their soil suitability from 1 to 9 (1 being very unsuitable to 9 being extremely suitable). This criterion is qualitative and positive. For the stochastic approach, a pert distribution was considered. In the deterministic approach, the mean value was used.

7. Levelized cost of electricity (LCoE): This criterion considers an estimation of the LCoE for each offshore location (2015 £/MWh). The values were calculated according to the DECC simple levelized cost of energy model in [59]. The calculations assumed an 8 MW size turbine. Jacket structure and a range of water depths (maximum and minimum water depth measured in each site) per offshore site. The criterion is quantitative and negative. In the stochastic approach, the triangle distribution was used and in the deterministic, the mean value.

This study considered the criteria that have greater impact than others in the final decision-making process by assigning weights derived from the insights of experts.

It should be noted that some aspects were excluded for this analysis as they do not appear to affect the location selection process or they already included in the existing selected criteria and other steps of the framework. Fisheries and aquaculture is a criterion that considers the positive effects of the aquaculture and the fisheries around the wind farms. The criterion could be assessed according to similar fisheries and aquacultures that seem to benefit from nearby wind farms. This information is hard to obtain systematically or does not meet the unique characteristics of the wind farm locations. Regarding the environmental extensions, such as birds and fish, these were not considered in the environmental impact criterion. The Department for Energy and Climate Change conducted a strategic environmental assessment on the offshore sites for over 60m of water depth around the UK and the Crown Estate identified possible suitable areas for offshore wind farm deployments aligned with government policy and released the 3 Rounds [11,13]. Further, service life extension will not be considered because of the nature of the problem. In order to consider life extension, a sample of individual turbines is monitored, tested and investigated. There is no evidence whether there is a link of life extension possibility to the offshore location. Finally, marine growth or artificial reefs will not be included in the study because it does not reveal the uniqueness of the offshore sites. Marine growth exists in all offshore structures.

4. Results and Discussion

The data obtained from the experts were analysed and used in MCDM both deterministically and stochastically. The results from all locations (from all five zones) are provided and illustrated in Figure 9 as cost breakdown analysis. All 7 solutions shown and discussed were obtained from the execution of the NSGA II, and they are equally optimal solutions, according to the Pareto equality. The problem considered all 18 sites from the five selected Round 3 zones and the optimum results minimize CAPEX, OPEX and $C_{D\&D}$, as shown in Figure 9. At the same time, the remaining objectives are also optimized. All layouts were found to deliver optimal solutions, where layout 3 was found only once with few turbines.

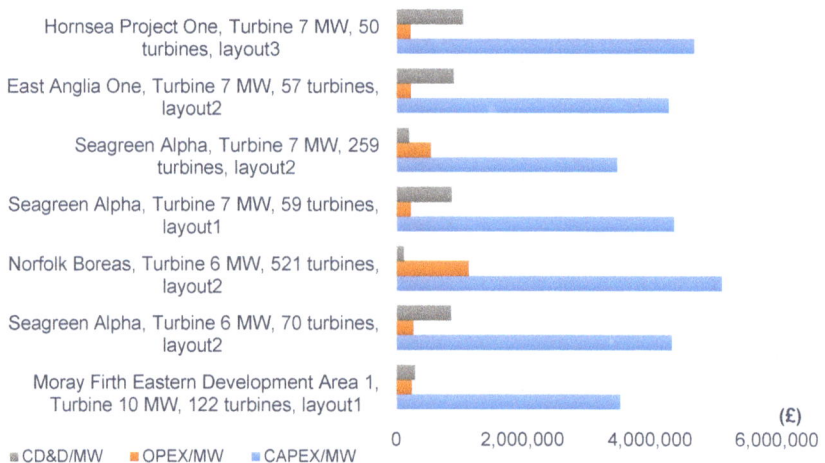

Figure 9. Cost breakdown per MW For all Pareto Front solutions for layout cases 1, 2 and 3.

All optimal solutions are listed in Table 5. The solution that includes Hornsea Project One and layout 3 delivered the lowest costs of the optimal solutions. Although that was expected as it was found that only 50 turbines were selected by the optimizer, the same solution is the second most expensive per MW as shown in Figure 9. Moray Firth Eastern Development Area 1 could deliver the lowest cost per MW. The three solutions of the Seagreen Alpha included both layouts 1 and 2. The fact that Seagreen Alpha was selected three times shows the flexibility of multiple options for a suitable budget assignment that the framework can deliver to the developers. The $C_{D\&D}$ presents low fluctuations for all solutions. In the range between £2 and £2.3 billion of the total cost, four solutions were discovered, for the areas of Seagreen Alpha (twice), East Anglia One and Hornsea Project One. Figure 10 illustrates the % frequency of the occurrences of the optimal solutions. Five locations were selected from the 18 in total. Seagreen Alpha was selected three times more than the rest of the optimum solutions.

Table 5. Numerical results for all zones.

Offshore Wind Farm Site	Layout Selected	Turbine Size (MW)	NWT	OPEX (£)	CD&D [£]	CAPEX [£]	Total Cost [£]
Moray Firth Eastern Development Area 1	layout 1	10	122	307,322,672	365,371,991	4,316,454,016	4,989,148,680
Seagreen Alpha	layout 2	6	70	115,563,086	365,329,300	1,821,862,415	2,302,754,802
Norfolk Boreas	layout 2	6	521	3,612,087,515	383,807,107	16,034,493,829	20,030,388,452
Seagreen Alpha	layout 1	7	59	97,590,070	363,801,519	1,806,818,815	2,268,210,405
Seagreen Alpha	layout 2	7	259	996,944,713	373,550,029	6,323,114,490	7,693,609,234
East Anglia One	layout 2	7	57	93,654,614	364,474,208	1,712,388,330	2,170,517,154
Hornsea Project One	layout 3	7	50	81,096,384	371,523,572	1,640,942,787	2,093,562,744

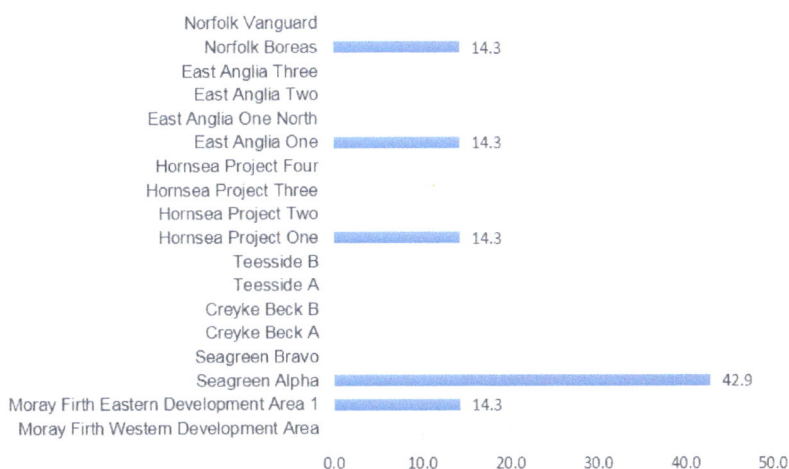

Figure 10. Percent of frequency of occurrences of optimal locations. Five sites were revealed by the optimizer.

The output of MOO is used as an input to the MCDM process. The output of TOPSIS is a prioritization of the alternatives (i.e., the five offshore sites). Two variations of TOPSIS (i.e., deterministic and stochastic) are employed. By combining those two methods, MOO and MCDM, the best location is identified, and the decision maker's confidence increases. These five locations were selected to take part in the MCDM process in order to be further discussed and to obtain a ranking of the locations using the stochastic expansion of TOPSIS. Following the process of TOPSIS, the considered alternatives are listed in Table 6, which are all considered to be unoccupied and available for a new wind farm installation for the purposes of the problem.

Table 7 shows the final decision matrix with the mean values for every alternative versus criterion. The criteria and alternatives' IDs were used for clarity and simplification. All qualitative inputs were scaled from 1 to 9, as mentioned before. Table 8 shows the frequency of the experts' preference per criterion and the normalized mean values of the weights extracted from them.

Table 6. List of alternatives.

Alternatives/Zones	Wind Farm Site Name	ID
Moray Firth	Moray Firth Eastern Development Area 1	A1
Firth of Forth	Seagreen Alpha	A2
Hornsea	Hornsea Project One	A3
East Anglia (Norfolk Bank)	East Anglia One	A4
East Anglia (Norfolk Bank)	Norfolk Boreas	A5

Table 7. Decision matrix.

Alternatives/Criteria	C1	C2	C3	C4	C5	C6	C7
A1	4.5	11.5	61,979,649,702	25.8	226	5.6	118.7
A2	4.5	10.4	31,984,700,386	25.8	157.5	6.4	129.2
A3	7	10.0	65,153,119,337	26.0	5939	6.4	114.2
A4	6	9.8	29,122,509,239	25.8	1859	6.7	114.5
A5	4.5	10.0	39,619,870,326	25.8	1859	6.7	114.2

Table 8. Frequency of experts' preference per criterion.

Rate (1–5)	Criteria						
	C1	C2	C3	C4	C5	C6	C7
1 Not at all important	0	0	0	0	0	0	0
2. Slightly important	1	1	5	1	1	0	1
3. Moderately important	5	1	3	6	2	5	1
4. Very important	4	7	2	2	6	7	4
5. Extremely important	3	4	3	4	4	1	7
Normalized mean weights	0.138	0.153	0.121	0.138	0.150	0.138	0.161

Specifically for the calculation of C6 against alternatives in Table 7, input from three experts was considered. Although the number of experts replying to the seven criteria was mentioned before (i.e., 13), a different number of experts (i.e., 3) was involved in the estimation of the geotechnical condition criterion in order to form the distribution from their answers. The reason that the number of experts was not the same in the two procedures is that different expertise was required in both cases. The geotechnical conditions can be better perceived by geotechnical engineers, and the total number of experts is very specific and more difficult to engage with. Based on experts' answers, the normalized mean weights of the criteria are estimated by the frequency of experts' preferences per criterion in Table 8.

The results of both variations of TOPSIS are listed in Table 9. By implementation, the stochastic variation reveals more quantitative information about the alternatives and assigns the probability that an option will rank first, as shown in Figure 11. According to stochastic TOPSIS, the alternative that involves Seagreen Alpha was the most probable solution, followed by Moray Firth Eastern Development Area 1. Also, the former is three times more probable to be selected compared to the latter. The probability of other options to be selected is significantly lower, and Hornsea Project One is unlikely to be selected.

Table 9. Results of deterministic and stochastic Technique for the Order of Preference by Similarity to the Ideal Solution (TOPSIS).

Alternatives	Deterministic TOPSIS		Stochastic TOPSIS	
	Score	Rank	Score	Rank
A1	0.733	2	21.88%	2
A2	0.816	1	64.44%	1
A3	0.181	5	0.00%	5
A4	0.712	3	10.22%	3
A5	0.660	4	3.50%	4

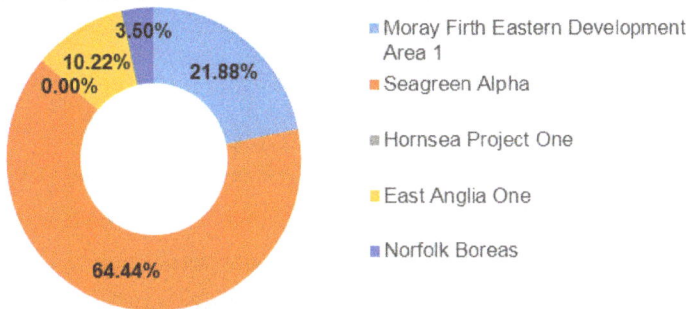

Figure 11. Probability chart of the stochastic TOPSIS.

In the survey, the experts were asked to make recommendations or leave comments about the criteria in order to include their insight in future studies or the limitations section. As expected, most experts made some recommendations that are worth considering in the next steps. Some experts responded according to their understanding of the work that is carried out and the work that was done before this study. Some of them pointed out factors that were already included in the study in the modelling of the work or already included in the criteria given to them, for example, the grid availability and the power prices.

The importance of the operational environmental conditions was pointed out and how critical they think it is as it drives the wind farm's maximum output and capacity factor. It was also stated that the wind speed should be taken into account separately in the study. The geotechnical conditions and the soil's impact on the design (both substructure and transmission system) were also pointed out. One expert made clear this should not be overlooked. The geotechnical conditions were studied separately and finally incorporated into this study as explained above.

At the end of the survey, the experts were asked to include any other criteria that can affect the location selection. One suggestion was to include the consenting process as it can be affected by environmental reasons such as the protection of biodiversity. This problem was seen in a wind farm due to Sabellaria reefs in the past. The ease and time of consent were also raised by another expert. It was suggested that specific stakeholders should be asked to participate such as the Ministry of Defence, air traffic, shipping, fishing, etc.

The government support mechanism came up in the comments a few times. It was also mentioned that the government regulations for each location need to be checked, because in many cases it might be a better decision to open the market in other continents. Also, the project financing and other contracts for difference (CfD) opportunities were mentioned. On top of that, the access to human resources was pointed out to show the impact of different locations.

Also, it was mentioned that if floating support structures were considered in the study, then the water depth and availability of relatively large and deep shipyards would be very important constraints. In this case, floating structures were not considered, but they could be included in the future.

The results of the study could also impact the criteria and the way these locations are selected by the Crown Estate providing more informed and cost-efficient options for future developers. Considerable actions are mandated on top of the development plans for minimizing investment, developing the supply chain, securing consents, ensuring economic grid investment and connection, and accessing finance [2,5].

5. Conclusions

The coupling of MOO with MCDM and expert surveys was demonstrated in this paper, as a method to increase the confidence of wind energy developers at the early stages of the investment. A set of locations from Round 3 and types of turbines were considered in the LCC analysis. By employing NSGAII and two variations of TOPSIS, optimum solutions were revealed and ranked based on experts' preferences. In the current problem formulation, among the optimum solutions, Seagreen Alpha was the best option, and Hornsea Project One was the least probable to be selected. From the surveys, additional criteria and stakeholders were recommended by the participants, which will be considered in the future.

The proposed methodology could also be applied to other sectors in order to increase investment confidence and provide optimum solutions. For example, the installation of floating offshore wind and wave devices could benefit from the framework where the optimum locations can be suggested concerning cost and operational aspects of each technological need. The foundation in this study is considered to be the jacket structure because the LCC is formulated accordingly. More LCC parametric analyses can be investigated in the future for different types of structures.

Author Contributions: V.M. carried out the research and documented the findings. E.L.-M., as an associate, provided domain expertise in the scientific field of Multi-Criteria Decision Making and guidance in the implementation of the related processes. A.K. provided overall guidance and quality assurance in the publication.

Funding: This work was supported by Grant EP/L016303/1 for Cranfield University, Centre for Doctoral Training in Renewable Energy Marine Structures (REMS) (http://www.rems-cdt.ac.uk/) from the UK Engineering and Physical Sciences Research Council (EPSRC). Data underlying this paper can be accessed at https://doi.org/10. 17862/cranfield.rd.6292703.

Conflicts of Interest: The authors declare that there is no conflict of interest.

References

1. Flood, D. *Round 3 Offshore Wind Farms*. *UK Future Energy Scenarios Seminar*; National Grid: Warwick, UK, 2012.
2. Department of Energy and Climate Change. *UK Renewable Energy Roadmap*; Department of Energy and Climate Change: London, UK, 2011.
3. Weston, D. Offshore Wind Costs Fall 32% Since 2011. Available online: https://www.windpoweroffshore. com/article/1421825/offshore-wind-costs-fall-32-2011 (accessed on 13 July 2018).
4. Reuters. Denmark's Dong Wins UK Contract to Build World's Largest Offshore Wind Farm. Available online: https://uk.reuters.com/article/uk-britain-renewables-dong-energy/denmarks-dong-wins-uk-contract-to-build-worlds-largest-offshore-wind-farm-idUKKCN1BM0R1 (accessed on 13 July 2018).
5. Renewables First. What Are the Main Project Risks for Wind Power? Available online: https://www.renewablesfirst.co.uk/windpower/windpower-learning-centre/what-are-the-main-project-risks-for-wind-power/ (accessed on 13 July 2018).
6. HM Government. *Offshore Wind Industrial Strategy Business and Government Action*; BIS: London, UK, 2013.
7. Mytilinou, V.; Kolios, A.J.; Di Lorenzo, G. A comparative multi-disciplinary policy review in wind energy developments in europe. *Int. J. Sustain. Energy* **2015**, *36*, 1–21. [CrossRef]

8.   European Observation Network for Territorial Development and Cohesion. Inspire Policy Making by Territorial Evidence. Available online: http://www.espon.eu/main/Menu_Publications/Menu_MapsOfTheMonth/map1101.html (accessed on 28 November 2017).

9.   Lin, S.-Y.; Chen, J.-F. Distributed optimal power flow for smart grid transmission system with renewable energy sources. *Energy* **2013**, *56*, 184–192. [CrossRef]

10. Burton, T.; Jenkins, N.; Sharpe, D.; Bossanyi, E. *Wind Energy Handbook*, 2nd ed.; Wiley: Hoboken, NJ, USA, 2011.

11. The Crown Estate. *Round 3 Zone Appraisal and Planning. A Strategic Approach to Zone Design, Project Identification and Consent*; The Crown Estate: London, UK, 2010.

12. The Crown Estate. The Crown Estate Announces Round 3 Offshore Wind Development Partners. Available online: http://www.thecrownestate.co.uk/news-and-media/news/2010/the-crown-estate-announces-round-3-offshore-wind-development-partners/ (accessed on 28 October 2017).

13. The Crown Estate. *Round 3 Offshore Wind Site Selection at National and Project Levels*; The Crown Estate: London, UK, 2013.

14. National Grid. *The Crown Estate. Round 3 Offshore Wind Farm Connection Study*; National Grid: London, UK, 2012.

15. Jones, D.F.; Wall, G. An extended goal programming model for site selection in the offshore wind farm sector. *Ann. Oper. Res.* **2016**, *245*, 121–135. [CrossRef]

16. Karimi, M.; Hall, M.; Buckham, B.; Crawford, C. A multi-objective design optimization approach for floating offshore wind turbine support structures. *J. Ocean Eng. Mar. Energy* **2017**, *3*, 69–87. [CrossRef]

17. Malekpoor, H.; Chalvatzis, K.; Mishra, N.; Mehlawat, M.K.; Zafirakis, D.; Song, M. Integrated grey relational analysis and multi objective grey linear programming for sustainable electricity generation planning. *Ann. Oper. Res.* **2017**. [CrossRef]

18. Gentils, T.; Wang, L.; Kolios, A. Integrated structural optimisation of offshore wind turbine support structures based on finite element analysis and genetic algorithm. *Appl. Energy* **2017**, *199*, 187–204. [CrossRef]

19. Fuller, S.K.; Petersen, S.R. Life-Cycle Costing Manual for the Federal Energy Management Program, 1995 Edition. NIST Handbook 135. Available online: https://ws680.nist.gov/publication/get_pdf.cfm?pub_id=907459 (accessed on July 2018).

20. Shafiee, M.; Brennan, F.; Espinosa, I.A. Whole life-cycle costing of large-scale offshore wind farms. In Proceedings of the Conference: European Wind Energy Association (EWEA), Paris, France, 17–30 November 2015.

21. Nordahl, M. *The Development of a Life Cycle Cost Model for an Offshore Wind Farm*; Chalmers University of Technology: Göteborg, Sweden, 2011.

22. Luengo, M.; Kolios, A. Failure mode identification and end of life scenarios of offshore wind turbines: A review. *Energies* **2015**, *8*, 8339–8354. [CrossRef]

23. Lozano-Minguez, E.; Kolios, A.J.; Brennan, F.P. Multi-criteria assessment of offshore wind turbine support structures. *Renew. Energy* **2011**, *36*, 2831–2837. [CrossRef]

24. Kolios, A.; Mytilinou, V.; Lozano-Minguez, E.; Salonitis, K. A comparative study of multiple-criteria decision-making methods under stochastic inputs. *Energies* **2016**, *9*, 566. [CrossRef]

25. Kolios, A.; Read, G.; Ioannou, A. Application of multi-criteria decision-making to risk prioritisation in tidal energy developments. *Int. J. Sustain. Energy* **2014**, *35*, 1–16. [CrossRef]

26. Kolios, A.J.; Rodriguez-Tsouroukdissian, A.; Salonitis, K. Multi-criteria decision analysis of offshore wind turbines support structures under stochastic inputs. *Ships Offshore Struct.* **2014**, *11*, 38–49.

27. Corrente, S.; Greco, S.; Kadziński, M.; Słowiński, R. Robust ordinal regression in preference learning and ranking. *Mach. Learn.* **2013**, *93*, 381–422. [CrossRef]

28. Angilella, S.; Corrente, S.; Greco, S.; Słowiński, R. Multiple criteria hierarchy process for the choquet integral. In *Evolutionary Multi-Criterion Optimization: Proceedings of the 7th International Conference, EMO 2013, Sheffield, UK, 19–22 March 2013*; Purshouse, R.C., Fleming, P.J., Fonseca, C.M., Greco, S., Shaw, J., Eds.; Springer: Berlin/Heidelberg, Germany, 2013; pp. 475–489.

29. Corrente, S.; Greco, S.; Słowiński, R. Multiple criteria hierarchy process in robust ordinal regression. *Decis. Support Syst.* **2012**, *53*, 660–674. [CrossRef]

30. Corrente, S.; Figueira, J.R.; Greco, S.; Słowiński, R. A robust ranking method extending electre III to hierarchy of interacting criteria, imprecise weights and stochastic analysis. *Omega* **2017**, *73*, 1–17. [CrossRef]

31. Corrente, S.; Doumpos, M.; Greco, S.; Słowiński, R.; Zopounidis, C. Multiple criteria hierarchy process for sorting problems based on ordinal regression with additive value functions. *Ann. Oper. Res.* **2017**, *251*, 117–139. [CrossRef]

32. Mateo, J.R.S.C. *Multi-Criteria Analysis in the Renewable Energy Industry*; Springer: Berlin/Heidelberg, Germany, 2012.

33. Martin, H.; Spano, G.; Küster, J.F.; Collu, M.; Kolios, A.J. Application and extension of the topsis method for the assessment of floating offshore wind turbine spport structures. *Ships Offshore Struct.* **2013**, *8*, 477–487. [CrossRef]

34. Perera, A.T.D.; Attalage, R.A.; Perera, K.K.C.K.; Dassanayake, V.P.C. A hybrid tool to combine multi-objective optimization and multi-criterion decision making in designing standalone hybrid energy systems. *Appl. Energy* **2013**, *107*, 412–425. [CrossRef]

35. Xiong, W.; Qi, H. A extended topsis method for the stochastic multi-criteria decision making problem through interval estimation. In Proceedings of the 2010 2nd International Workshop on Intelligent Systems and Applications, Wuhan, China, 22–23 May 2010.

36. Collu, M.; Kolios, A.; Chahardehi, A.; Brennan, F.; Patel, M.H. A multi-criteria decision making method to compare available support structures for offshore wind turbines. In Proceedings of the European Wind Energy Conference and Exhibition, Warsaw, Poland, 20–23 April 2010.

37. Malekpoor, H.; Chalvatzis, K.; Mishra, N.; Ramudhin, A. A hybrid approach of vikor and bi-objective integer linear programming for electrification planning in a disaster relief camp. *Ann. Oper. Res.* **2018**. [CrossRef]

38. Chalvatzis, K.J.; Malekpoor, H.; Mishra, N.; Lettice, F.; Choudhary, S. Sustainable resource allocation for power generation: The role of big data in enabling interindustry architectural innovation. *Technol. Forecast. Soc. Chang.* **2018**, in press. [CrossRef]

39. Lei, X.; Shiyun, T.; Yanfei, D.; Yuan, Y. Sustainable operation-oriented investment risk evaluation and optimization for renewable energy project: A case study of wind power in China. *Ann. Oper. Res.* **2018**. [CrossRef]

40. Mytilinou, V.; Kolios, A.J. A multi-objective optimisation approach applied to offshore wind farm location selection. *J. Ocean Eng. Mar. Energy* **2017**, *3*, 265–284. [CrossRef]

41. Espinosa, I.A. Life Cycle Costing of Offshore Wind Turbines. Master's Thesis, Cranfield University, Bedford, UK, 2014.

42. Dicorato, M.; Forte, G.; Pisani, M.; Trovato, M. Guidelines for assessment of investment cost for offshore wind generation. *Renew. Energy* **2011**, *36*, 2043–2051. [CrossRef]

43. Laura, C.-S.; Vicente, D.-C. Life-cycle cost analysis of floating offshore wind farms. *Renew. Energy* **2014**, *66*, 41–48. [CrossRef]

44. Wind Energy The facts. Development and Investment Costs of Offshore Wind Power. Available online: http://www.wind-energy-the-facts.org/development-and-investment-costs-of-offshore-wind-power.html (accessed on 28 October 2017).

45. 4Coffshore. Global Offshore Wind Farms Database. Available online: http://www.4coffshore.com/offshorewind/index.html?lat=50.668&lon=-0.275&wfid=UK36 (accessed on 28 October 2017).

46. The Crown Estate. *A Guide to an Offshore Wind Farm*; The Crown Estate: London, UK, 2017.

47. QGIS. A Free and Open Source Geographic Information System. Available online: http://www.qgis.org/en/site/index.html (accessed on 28 October 2017).

48. Department of Energy and Climate Change. *UK Ports for the Offshore Wind Industry: Time to Act*; Department of Energy and Climate Change: London, UK, 2009.

49. Marine Traffic. Available online: http://www.marinetraffic.com (accessed on 28 October 2017).

50. UK Ports Directory Ports. The Comprehensive Guide to the UK's Commercial Ports. Available online: http://uk-ports.org/ (accessed on 28 October 2017).

51. The Crown Estate. Maps and Gis Data. Available online: https://www.thecrownestate.co.uk/rural-and-coastal/coastal/downloads/maps-and-gis-data/ (accessed on 28 October 2017).

52. Samorani, M. The wind farm layout optimization problem. In *Handbook of Wind Power Systems*; Pardalos, P.M., Rebennack, S., Pereira, M.V.F., Iliadis, N.A., Pappu, V., Eds.; Springer: Berlin/Heidelberg, Germany, 2013; pp. 21–38.

53. Fugro GEOS. *Wind and Wave Frequency Distributions for Sites Around the British Isles*; Fugro GEOS: Southampton, UK, 2001.

54. Hadka, D. Platypus. Available online: http://platypus.readthedocs.io/en/latest/experimenter.html# comparing-algorithms-visually (accessed on 28 October 2017).

55. Hwang, C.; Yoon, K. Multiple attribute decision making. In *Lecture Notes in Economics and Mathematical Systems*; Springer: Berlin/Heidelberg, Germany, 1981.

56. Yoon, K.P.; Hwang, C.-L. *Multiple Attribute Decision Making: An Introduction*; Sage Publications: Thousand Oaks, CA, USA, 1995; Volume 104.

57. Zeleny, M. Multiple Criteria Decision Making (MCDM): From Paradigm Lost to Paradigm Regained? *J. Multi-Crit. Decis. Anal.* **2011**, *18*, 77–89. [CrossRef]

58. 4COffshore. Offshore Turbine Database. Available online: http://www.4coffshore.com/windfarms/ turbines.aspx (accessed on 28 October 2017).

59. BVG Associates. DECC Offshore Wind Programme—Simple Levelised Cost of Energy Model Revision 3–26/10/2015. 2015. Available online: http://www.demowind.eu/LCOE.xlsx (accessed on 1 November 2017).

Article

A Stochastic Inexact Robust Model for Regional Energy System Management and Emission Reduction Potential Analysis—A Case Study of Zibo City, China

Yulei Xie [1],*, Linrui Wang [2], Guohe Huang [3], Dehong Xia [1] and Ling Ji [4]

[1] School of Energy and Environmental Engineering, University of Science and Technology Beijing, Beijing 100083, China; xia@me.ustb.edu.cn

[2] The Vehicle Pollution Prevention and Control Center of Jinan, Jinan Environmental Protection Bureau, Jinan 250099, Shandong, China; jnepbstc@jnep.cn

[3] Environmental Systems Engineering Program, Faculty of Engineering, University of Regina, Regina, SK S4S 0A2, Canada; gordon.huang@uregina.ca

[4] School of Economics and Management, Beijing University of Technology, Beijing 100124, China; hdjiling@bjut.edu.cn

* Correspondence: xieyulei001@ustb.edu.cn; Tel.: +86-106-233-4971; Fax: +86-106-232-9145

Received: 4 July 2018; Accepted: 3 August 2018; Published: 13 August 2018

Abstract: In this study, in order to improve regional energy system adjustment, a multistage stochastic inexact robust programming (MSIRP) is proposed for electric-power generation planning and structure adjustment management under uncertainty. Scenario-based inexact multistage stochastic programming and stochastic robust optimization were integrated into general programming to reflect uncertainties that were expressed as interval values and probability distributions in the objective function and constraints. An MSIRP-based energy system optimization model is proposed for electric-power structure management of Zibo City in Shandong Province, China. Three power demand scenarios associated with electric-power structure adjustment, imported electricity, and emission reduction were designed to obtain multiple decision schemes for supporting regional sustainable energy system development. The power generation schemes, imported electricity, and emissions of CO_2 and air pollutants were analyzed. The results indicated that the model can effectively not only provide a more stable energy supply strategies and electric-power structure adjustment schemes, but also improve the balanced development between conventional and new clear power generation technologies under uncertainty.

Keywords: scenario-based multistage stochastic programming; energy system management model; stochastic robust optimization; electric-power structure adjustment; energy conservation and emissions reduction

1. Introduction

Rapid power consumption increment, increasing deterioration of environmental quality, and imperfect energy system management have led to unsustainable energy resources exploitation and utilization, unreasonable electric-power structure, and serious environmental issues [1–3]. In order to search effective and suitable energy development strategies for regional condition, energy system management and planning has become a priority for many countries and regions. However, multiple forms of uncertain information are involved in energy system management and the related social-economic factors and/or technical-economic parameters, causing a variety of complexities in decision support and policy analysis for regional energy planning [4]. In addition, such complexities would pose great challenges in formulating more scientific and reasonable development strategies for decision-makers, and have serious impact on the effectiveness of energy supply schemes. Therefore, it is desirable to develop effective uncertain optimization models/techniques for energy system management and the related decision analysis.

Previously, a great number of inexact programming approaches were proposed for helping energy system planning and management in different regional scales [5–14]. For example, Cai et al. (2009) advanced an interval parameter interactive decision support system for energy system management under reflecting uncertainties as interval values [15]. Li et al. (2010) proposed an inexact fuzzy multistage stochastic energy system management model for supporting regional electric-power generation and capacity planning, where interval parameter programming, mixed integer linear programming, multistage stochastic programming, and fuzzy linear programming were incorporated into a general optimization framework [16]. Li et al. (2011) proposed a fuzzy stochastic energy system optimization model associated with renewable energy development and greenhouse gas mitigation, where the uncertainties in the objective and constraints were expressed as fuzzy interval functions, interval values, and discrete probability distributions [17]. Huang et al. (2017) developed an inexact fuzzy stochastic chance-constrained programming for evacuation management of nuclear power plant, where interval parameter programming and fuzzy stochastic chance-constrained programming were integrated into a general framework for dealing with uncertainties [18]. Sheikhahmadi et al. (2018) proposed a risk-based two-stage stochastic programming for microgrid system operation management, where two-stage stochastic programming was to reflect uncertainties of renewable energy, and conditional value at risk index was used to avoid the system risk [19].

Among these methods, scenario-based interval multistage stochastic programming, as a hybrid method of interval parameter programming and scenario-based multistage stochastic programming, could deal with uncertainties presented as interval numbers and random distributions, and have been widely applied in energy system management [20–22]. For example, Xie et al. (2010) advanced an inexact fixed-mix multistage stochastic programming for long-term greenhouse gas emission reduction management in a regional scale energy system, where the fixed probability multistage stochastic programming and interval-parameter programming were integrated for expressing uncertainties in energy system management problems [23]. Wu et al. (2015) proposed an integrated method with interval-parameter programming, chance-constraint programming, and multistage stochastic programming for the coupled biomass–municipal solid waste power system operation management, which could reflect uncertainties as interval information and random distributions over a multistage context [24]. Golari et al. (2016) presented a production-inventory planning model in a multi-plant manufacturing system powered with onsite and grid renewable energy, where multistage stochastic programming was used to reflect system dynamic and uncertainties [25]. Fu et al. (2017) advanced an interval multistage fuzzy-stochastic programming for regional electric-power system management under considering environmental quality constraints, where interval-parameter programming, multistage stochastic programming, and fuzzy probability distribution was integrated to reflect the uncertain information and dynamic variation in the energy system [26]. Wang et al. (2018) developed multistage joint-probabilistic left-hand-side chance-constrained fractional programming for electric-power system planning considering climate change mitigation [27].

Although scenario-based inexact multistage stochastic programming had been successfully applied in many fields, it could not directly and effectively avoid the risk of random events, and the limitations would pose threats to system stability. Based on this point, stochastic robust optimization (SRO) is proposed for solving the problems through introducing the risk-aversion attitude into optimization models and obtaining robust solutions for stochastic system management [28–30]. The methods that coupled with the scenario-based inexact multistage stochastic programming and SRO have been used in solving many energy and environmental problems, such as water resources allocation, electric-power generation, and water/air quality management. For example, Chen et al. (2013) developed an interval robust-optimization model for CO_2 emission reduction management in energy systems, where the robustness measures were introduced to examine whether the second-stage cost variability could meet the expected levels or not [31]. Xie et al. (2014) proposed an inexact stochastic risk-aversion model for electric-power structure adjustment and pollutant emission management, where interval-parameter programming, stochastic robust optimization, and multistage stochastic programming were integrated to address system uncertainties [32].

Therefore, the aim of this study is to formulate a multistage stochastic inexact robust programming (MSIRP) model to support regional electric-power system management coupled with pollutant mitigation constraints and power structure adjustment requirements in Zibo City, China. The method could not only reflect multiple uncertainties expressed as interval values and probability distribution, but also make a tradeoff between system risk and cost according to the decision-makers' attitudes. The modeling results will be helpful for local decision-makers to choose cost–risk electric-power generation schemes, and obtain reasonable electric-power structure adjustment strategies. The rest structure organization of this paper is provided as follows. The development process and solution algorithm of multistage stochastic inexact robust programming (MSIRP) is introduced in Section 2. The overview of the energy system of Zibo City are described, and a MSIRP-based energy structure adjustment model is proposed in Section 3. The obtained results and deep discovery of the case study are analyzed and discussed in Section 4. The main conclusions are presented in Section 5.

2. Methodology

In regional energy systems, dynamic characters, discrete probability distributions, intervals information, and policy implications were addressed through scenario-based inexact multistage stochastic programming, and SRO could effectively handle the system risk. The modeling framework of the MSIRP could obtain applicable and reasonable solutions under different random scenarios corresponding to power generation targets for decision-makers in order to support the energy system development in the future.

2.1. Inexact Scenario-Based Multistage Stochastic Programming

In the scenario-based multistage stochastic programming, the probabilities $p_{tk}(t = 1, 2, \ldots, T; k = 1, 2, \ldots, K_t,)$ of the stochastic event have predefined values, and the parameters without probability can be reflected as interval values. Thus, the scenario-based inexact multistage stochastic programming can be expressed as follows [33,34]:

$$Min\ f^{\pm} = \sum_{t=1}^{T}\sum_{j=1}^{n_1} c_{jt}^{\pm}x_{jt}^{\pm} + \sum_{t=1}^{T}\sum_{j=1}^{n_1}\sum_{k=1}^{K_t} p_{tk}d_{jt}^{\pm}y_{jtk}^{\pm}, \tag{1}$$

subject to

$$\sum_{j=1}^{n_1} a_{rjt}^{\pm}x_{jt}^{\pm} \leq b_{rt}^{\pm}, \forall r, t \tag{2}$$

$$\sum_{j=1}^{n_1} a_{ijt}^{\pm}x_{jt}^{\pm} + \sum_{j=1}^{n_1} e_{ijt}^{\pm}y_{jkt}^{\pm} \leq \tilde{w}_{itk}^{\pm}, \forall i, t, k \tag{3}$$

$$x_{jt}^{\pm} \geq 0, \forall t, j = 1, 2, \ldots, n_1 \tag{4}$$

$$y_{jkt}^{\pm} \geq 0, \forall t, k, j = 1, 2, \ldots, n_1 \tag{5}$$

where p_{tk} is the probability for scenario k in period t; for each period t, the total number of scenarios is denoted as K_t, and $\sum_{k=1}^{K_t} p_{tk} = 1$; and $\tilde{w}_{ikt}^{\pm}$ represents the random parameter in the model associated with the occurrence probability p_{tk} in period t. $x_{jt}^{\pm}$ denotes the first-stage variables that have to be determined before the random event occurrence; and $y_{jkt}^{\pm}$ are the second-stage variables that have to be decided for making a recourse actions to fulfil validity of the decision-making after the random event occurrence.

2.2. Inexact Multistage Stochastic Robust Programming

The proposed inexact scenario-based multistage stochastic programming can effectively reflect stochastic information, interval values, and dynamic feature by means of discrete random variables in long-term planning problems. However, Model (1) could not effectively reflect the system risk introduced by random information, that directly affect the feasibility and reliability of the proposed model. SRO is an effective choice for solving such problems, and it can be introduced into Model (1), that leads to a multistage stochastic inexact robust programming (MSIRP) as follows [32]:

$$
\begin{aligned}
\text{Min } f^{\pm} =\ & \sum_{t=1}^{T}\sum_{j=1}^{n_1} c_{jt}^{\pm} x_{jt}^{\pm} + \sum_{t=1}^{T}\sum_{j=1}^{n_2}\sum_{k=1}^{K_t} p_{tk} d_{jt}^{\pm} y_{jtk}^{\pm} \\
& + \omega \sum_{t=1}^{T}\sum_{j=1}^{n_2}\sum_{k=1}^{K_t} p_{tk} \left| d_{jt}^{\pm} y_{jtk}^{\pm} - \sum_{j=1}^{n_2}\sum_{k=1}^{K_t} p_{tk} d_{jt}^{\pm} y_{jtk}^{\pm} \right|
\end{aligned}
\tag{6}
$$

subject to

$$
\sum_{j=1}^{n_1} a_{rjt}^{\pm} x_{jt}^{\pm} \le b_{rt}^{\pm}, \forall r, t
\tag{7}
$$

$$
\sum_{j=1}^{n_1} a_{ijt}^{\pm} x_{jt}^{\pm} + \sum_{j=1}^{n_1} e_{ijt}^{\pm} y_{jkt}^{\pm} \le \widetilde{w}_{itk}^{\pm}, \forall i, t, k
\tag{8}
$$

$$
x_{jt}^{\pm} \ge 0, \forall t, j = 1, 2, \ldots, n_1
\tag{9}
$$

$$
y_{jkt}^{\pm} \ge 0, \forall t, k, j = 1, 2, \ldots, n_1
\tag{10}
$$

where the non-negative factor ω denotes a trade-off weight coefficient; and $\left| d_{jt}^{\pm} y_{jtk}^{\pm} - \sum_{j=1}^{n_2}\sum_{k=1}^{K_t} p_{tk} d_{jt}^{\pm} y_{jtk}^{\pm} \right|$ is a variability measure for reflecting the multistage recourse costs. The objective of Model (6) is a nonlinear function, and according to [35,36], the model can be converted into a linear programming model as follows:

$$
\begin{aligned}
\text{Min } f^{\pm} =\ & \sum_{t=1}^{T}\sum_{j=1}^{n_1} c_{jt}^{\pm} x_{jt}^{\pm} + \sum_{t=1}^{T}\sum_{j=1}^{n_1}\sum_{k=1}^{K_t} p_{tk} d_{jt}^{\pm} y_{jtk}^{\pm} \\
& + \omega \sum_{t=1}^{T}\sum_{j=1}^{n_1}\sum_{k=1}^{K_t} p_{tk} \left(d_{jt}^{\pm} y_{jtk}^{\pm} - \sum_{j=1}^{n_1}\sum_{k=1}^{K_t} p_{tk} d_{jt}^{\pm} y_{jtk}^{\pm} + 2\theta_{jkt}^{\pm} \right)
\end{aligned}
\tag{11}
$$

subject to

$$
d_{jt}^{\pm} y_{jtk}^{\pm} - \sum_{j=1}^{n1}\sum_{k=1}^{K_t} p_{tk} d_{jt}^{\pm} y_{jtk}^{\pm} + \theta_{jkt}^{\pm} \ge 0, \forall k, j = 1, 2, \ldots, n_1
\tag{12}
$$

$$
\sum_{j=1}^{n_1} a_{rjt}^{\pm} x_{jt}^{\pm} \le b_{rt}^{\pm}, \forall r, t
\tag{13}
$$

$$
\sum_{j=1}^{n_1} a_{ijt}^{\pm} x_{jt}^{\pm} + \sum_{j=1}^{n_1} e_{ijt}^{\pm} y_{jkt}^{\pm} \le \widetilde{w}_{itk}^{\pm}, \forall i, t, k
\tag{14}
$$

$$
x_{jt}^{\pm} \ge 0, \forall t, j = 1, 2, \ldots, n_1
\tag{15}
$$

$$
y_{jkt}^{\pm} \ge 0, \forall t, k, j = 1, 2, \ldots, n_1
\tag{16}
$$

where, through introducing the slack variable $\theta_{jkt}^{\pm}$, the objective can be transferred into a linear function as well as generate a specific control constraint (12). For Model (11), the first-stage variables $x_{jt}^{\pm}$ are considered/inputted as interval values with the lower and upper bound, and this cannot be directly solved using the existing methods. In this study, let μ_{jt} be a decision variable, $\mu_{jt} \in [0,1]$;

$\Delta x_{jt} = x_{jt}^+ - x_{jt}^-$, the first-stage variable $x_{jt} = x_{jt}^- + \mu_{jt}\Delta x_{jt}$, and μ_{jt} are intermediate decision variables for obtaining an optimized target values of the first-stage to support the related policy analyses [32]. According to [37], the MSIRP model can be transformed into two linear submodels, and the submodel corresponding to f^- can be firstly transformed as follows (assume that $c_{jt}^\pm \geq 0$, $\hat{w}_{itk}^+ > 0$, $b_{rt}^\pm > 0$, and $f^\pm > 0$):

$$
\begin{aligned}
\text{Min } f^- = {} & \sum_{t=1}^{T}\sum_{j=1}^{n_1} c_{jt}^-(x_{jt}^- + \mu_{jt}\Delta x_{jt}) + \sum_{t=1}^{T}\sum_{k=1}^{K_t} p_{tk}\Big(\sum_{j=1}^{j_1} d_{jt}^- y_{jtk}^- + \sum_{j=j_1+1}^{n_1} d_{jt}^- y_{jtk}^+\Big) \\
& + \omega \sum_{t=1}^{T}\sum_{k=1}^{K_t} p_{tk}\Big[\sum_{j=1}^{j_1}(d_{jt}^- y_{jtk}^- - \sum_{j=1}^{j_1}\sum_{k=1}^{K_t} p_{tk} d_{jt}^- y_{jtk}^- + 2\theta_{jkt}^-)\Big] \\
& + \omega \sum_{t=1}^{T}\sum_{k=1}^{K_t} p_{tk}\Big[\sum_{j=j_1+1}^{j_1}(d_{jt}^- y_{jtk}^+ - \sum_{j=j_1+1}^{n_1}\sum_{k=1}^{K_t} p_{tk} d_{jt}^- y_{jtk}^+ + 2\theta_{jkt}^-)\Big]
\end{aligned}
\tag{17}
$$

subject to

$$
\begin{aligned}
& \sum_{j=1}^{j_1} \big|d_{jtk}\big|^- Sign(d_{jtk}^-) y_{jkt}^- + \sum_{j=j_1+1}^{n_1} \big|d_{jtk}\big|^- Sign(d_{jtk}^-) y_{jkt}^+ \\
& - \sum_{j=1}^{j_1}\sum_{k=1}^{K_t} p_{tk}\big|d_{jtk}\big|^- Sign(d_{jtk}^-) y_{jkt}^- - \sum_{j=1}^{n_1}\sum_{k=1}^{K_t} p_{tk}\big|d_{jtk}\big|^- Sign(d_{jtk}^-) y_{jkt}^+ + \theta_{jtk}^- \geq 0, \forall i, j
\end{aligned}
\tag{18}
$$

$$
\sum_{j=1}^{n_1} \big|a_{rjt}\big|^+ Sign(a_{rjt}^+)(x_{jt}^- + \mu_{jt}\Delta x_{jt}) \leq b_{rt}^-, \forall r, t
\tag{19}
$$

$$
\begin{aligned}
& \sum_{j=1}^{n_1} \big|a_{ijt}\big|^+ Sign(a_{rjt}^+)(x_{jt}^- + \mu_{jt}\Delta x_{jt}) + \sum_{j=1}^{j_1} \big|e_{ijt}\big|^+ Sign(e_{ijt}^+) y_{jkt}^- \\
& + \sum_{j=j_1+1}^{n_1} \big|e_{ijt}\big|^- Sign(e_{ijt}^-) y_{jkt}^+ \leq \tilde{w}_{itk}^-, \forall i, t, k
\end{aligned}
\tag{20}
$$

$$
x_{jt}^- + \mu_{jt}\Delta x_{jt} \geq 0, \forall j, t
\tag{21}
$$

$$
0 \leq \mu_{jt} \leq 1, \forall j, t
\tag{22}
$$

$$
y_{jkt}^- \geq 0, \forall t, k, j = 1, 2, \ldots, j_1
\tag{23}
$$

$$
y_{jkt}^+ \geq 0, \forall t, k, j = j_1 + 1, j_1 + 2, \ldots, n_1
\tag{24}
$$

where μ_{jt}, $y_{jkt}^-(j = 1, 2, \ldots, j_1)$ and $y_{jkt}^+(j = j_1 + 1, j_1 + 2, \ldots, n_1)$ are the decision variables of model (17); $y_{jkt}^-(j = 1, 2, \ldots, j_1)$ and $y_{jkt}^+(j = j_1 + 1, j_1 + 2, \ldots, n_1)$ are the second-stage decision variables with positive and negative coefficients in the objective function; and the optimized solution of the first-stage variables are $x_{jtopt} = x_{jt}^- + \mu_{jtopt}\Delta x_{jt}$. Then, the submodel corresponding to f^+ can be expressed as follows:

$$
\begin{aligned}
\text{Min } f^+ = {} & \sum_{t=1}^{T}\sum_{j=1}^{n_1} c_{jt}^+ x_{jtopt} + \sum_{t=1}^{T}\sum_{k=1}^{K_t} p_{tk}\Big(\sum_{j=1}^{j_1} d_{jt}^+ y_{jtk}^+ + \sum_{j=j_1+1}^{n_1} d_{jt}^+ y_{jtk}^-\Big) \\
& + \omega \sum_{t=1}^{T}\sum_{k=1}^{K_t} p_{tk}\Big[\sum_{j=1}^{j_1}(d_{jt}^+ y_{jtk}^+ - \sum_{j=1}^{n_2}\sum_{k=1}^{K_t} p_{tk} d_{jt}^+ y_{jtk}^+ + 2\theta_{jkt}^+)\Big] \\
& + \omega \sum_{t=1}^{T}\sum_{k=1}^{K_t} p_{tk}\Big[\sum_{j=j_1+1}^{n_1}(d_{jt}^+ y_{jtk}^- - \sum_{j=j_1+1}^{n_1}\sum_{k=1}^{K_t} p_{tk} d_{jt}^+ y_{jtk}^- + 2\theta_{jkt}^+)\Big]
\end{aligned}
\tag{25}
$$

subject to

$$\sum_{j=1}^{j_1} \left|d_{jtk}\right|^+ Sign(d_{jtk}^+)y_{jkt}^+ + \sum_{j=j_1+1}^{n_1} \left|d_{jtk}\right|^+ Sign(d_{jtk}^+)y_{jkt}^-$$
$$-\sum_{j=1}^{j_1}\sum_{k=1}^{K_t} p_{tk}\left|d_{jtk}\right|^+ Sign(d_{jtk}^+)y_{jkt}^+ - \sum_{j=j_1+1}^{n_1}\sum_{k=1}^{K_t} p_{tk}\left|d_{jtk}\right|^+ Sign(d_{jtk}^+)y_{jkt}^- + \theta_{jtk}^+ \geq 0, \forall i,j \tag{26}$$

$$\sum_{j=1}^{n_1} \left|a_{rjt}\right|^- Sign(a_{rjt}^-)x_{jtopt} \leq b_{rt}^+, \forall r, t \tag{27}$$

$$\sum_{j=1}^{n_1} \left|a_{ijt}\right|^- Sign(a_{rjt}^-)\Delta x_{jtopt} + \sum_{j=1}^{j_1} \left|e_{ijt}\right|^- Sign(e_{ijt}^-)y_{jkt}^+ + \sum_{j=j_1+1}^{n_1} \left|e_{ijt}\right|^+ Sign(e_{ijt}^+)y_{jkt}^- \leq \tilde{w}_{itk}^+, \forall i, t, k \tag{28}$$

$$y_{jkt}^+ \geq y_{jktopt}^-, \forall t, k, j = 1, 2, \ldots, j_1 \tag{29}$$

$$y_{jktopt}^+ \geq y_{jkt}^- \geq 0, \forall t, k, j = j_1+1, j_1+2, \ldots, n_1 \tag{30}$$

where $y_{jkt}^+ (j = 1, 2, \ldots, j_1)$ and $y_{jkt}^- (j = j_1+1, j_1+2, \ldots, n_1)$ are decision variables that can be obtained through solving Submodel (25). Thus, the optimal solutions of Model (11) can be expressed as follows:

$$x_{jtopt} = x_{jt}^- + \mu_{jtopt}\Delta x_{jt} \tag{31}$$

$$y_{jktopt}^{\pm} = [y_{jktopt}^-, y_{jktopt}^+] \tag{32}$$

$$f_{opt}^{\pm} = [f_{opt}^-, f_{opt}^+]. \tag{33}$$

3. Case Study

3.1. Overview of Energy System in Zibo City

Zibo City ($35°55'20''\sim37°17'14''$ N, $117°32'15''\sim118°31'00''$ E), as shown in Figure 1, is located in the middle of Shandong province, China. Zibo City governs Zhangdian district, Zichuan district, Boshan district, Zhoucun district, Linzi district, Huantai country, Gaoqing country, and Yiyuan country, with a total area of 5938 km^2 and a total population of 4.61 million in 2014 [38]. In Zibo City, the manufacturing industry plays a significant role in supporting regional economic development; especially the ceramics manufacturing industry is famous around the world. For example, in 2014, the income of ceramic industry reached 112.8 billion yuan. In addition, high-new-technology industries (e.g., new materials, fine chemicals, and biological medicines) and other traditional industries (e.g., petrochemical industry, pharmaceuticals, metallurgy, and machinery and textiles) are developing rapidly in recent years. Moreover, in 2014, gross agricultural product reached to RMB 25.22 billion yuan, and the tertiary industry increased by RMB 163.45 billion yuan compared with 2013. In general, the rapid social-economic development is closely related with a higher power consumption. According to regional energy system statistic data in recent years, local electric-power generation is far from satisfying increasing regional demands.

Generally, the main electricity generation in Zibo City mainly relies on coal-fired power. The cogeneration power plays a large proportion in all electricity generation in Zibo, which could not only meet the demand of the district heating, but also greatly improve efficiency of coal resource utilization. In order to meet environmental requirements, there would have to be total consumption control on coal resources, according to the regional development plan from Zibo Municipal Development and Reform Committee. In addition, Zibo is abundant in renewable energy resources, such as solar, biomass, and wind, that have been considered as the primary options for addressing the crisis of electric-power shortage, and air pollutant and greenhouse gas mitigation. For instance, the average annual sunshine time reaches up to 2542.6 h with a greater potential and space for solar power and heat utilization. Moreover, throughout the windy corridor, in the surrounding of Boshan District and the southern mountain areas of Zichuan District, Zibo possesses the excellent conditions to build wind farms. According to regional energy development strategy of Zibo City (2010–2020), a greater number of renewable energy development plans have been promoted for adjusting the existing electric-power system structure, including 114 MW, 50 MW, and 244.5 MW of biomass and garbage power, solar, and wind power generation capacity by 2015, respectively. As a result, it will be helpful for alleviating the contradiction between energy supply capacity and consumption demands, and reducing atmospheric pollutants and carbon emission.

Figure 1. Location of the study area and regional energy resources distribution.

Although renewable energy has achieved development, and the government has also made great efforts to change regional electric-power structure, it still faces many challenges in electric-power system management. As a result of regional economic development, urbanization advance, and population growth, electric-power consumption and environmental quality requirement would be increasingly prominent, leading to an urgent need for regional electric-power structure adjustment. In this study, an inexact regional electric-power system optimization model is developed through multistage stochastic inexact robust programming for solving the following questions: (1) how to develop electric-power generation schemes for different power conversion technologies under air pollution and carbon mitigation requirements; (2) how to plan the overall development of renewable power conversion technologies and the proportion of imported electricity; (3) how to formulate more reasonable decision alternatives for decision-makers under different trade-offs between system cost and risk.

3.2. Electric-Power System Optimization Model Formulation

The developed multistage stochastic inexact robust programming is considered for regional electric-power system management in Zibo City. The objective is to achieve the optimal plans of electric-power supply with minimized system costs. The renewable power generation development, capacity expansion, and air pollutant and carbon emission reduction were also considered. Thus, the optimized model can be developed as follows:

$$Min \ f^{\pm} = f_1^{\pm} + f_2^{\pm} + f_3^{\pm} + f_4^{\pm} + f_5^{\pm} + f_6^{\pm} - f_7^{\pm} + f_8^{\pm} \tag{34}$$

[Costs for energy resources consumption]

$$f_1^{\pm} = \sum_{p=1}^{P} \sum_{t=1}^{T} PEC_{pt}^{\pm} \cdot (AE_{pt}^{\pm} + p_{th} \cdot DE_{pth}^{\pm}) \cdot EF_{pt}^{\pm} \tag{35}$$

[Costs for power generation]

$$f_2^{\pm} = \sum_{p=1}^{P} \sum_{t=1}^{T} PV_{pt}^{\pm} \cdot AE_{pt}^{\pm} + \sum_{p=1}^{P} \sum_{t=1}^{T} \sum_{h=1}^{H} p_{th} \cdot \left(PV_{pt}^{\pm} + PP_{pt}^{\pm}\right) \cdot DE_{pth}^{\pm} \tag{36}$$

[Cost for the district heating]

$$f_3^{\pm} = \sum_{p=1}^{P} \sum_{t=1}^{T} CV_{pt}^{\pm} \cdot (AH_{pt}^{\pm} + DH_{pt}^{\pm}) \tag{37}$$

[Costs for the expansion of installed capacity]

$$f_4^{\pm} = \sum_{p=1}^{P} \sum_{t=1}^{T} \sum_{h=1}^{H} p_{th} \cdot \left(YEH_{pth}^{\pm} \cdot A_{pt}^{\pm} + XEH_{pth}^{\pm} \cdot B_{pt}^{\pm}\right) \tag{38}$$

[Costs for atmospheric pollutants treatment]

$$f_5^{\pm} = \sum_{i=1}^{I} \sum_{p=1}^{P} \sum_{t=1}^{T} AE_{pt}^{\pm} \cdot \varsigma_{ipt}^{\pm} \cdot \left(1 - \eta_{ipt}^{\pm}\right) \cdot CPC_{it}^{\pm} + \sum_{i=1}^{I} \sum_{p=1}^{P} \sum_{t=1}^{T} \sum_{h=1}^{H} p_{th} \cdot DE_{pth}^{\pm} \cdot \varsigma_{ipt}^{\pm} \cdot \left(1 - \eta_{ipt}^{\pm}\right) \cdot DPC_{it}^{\pm} \tag{39}$$

[Costs for imported electric power]

$$f_6^{\pm} = \sum_{t=1}^{T} p_{th} \cdot IE_{th}^{\pm} \cdot IPE_t^{\pm} \tag{40}$$

[Subsidies for renewable energy generation]

$$f_7^{\pm} = \sum_{p=3}^{P} \sum_{t=1}^{T} \left(AE_{pt}^{\pm} + p_{th} \cdot DE_{pth}^{\pm}\right) \cdot SU_{pt}^{\pm} \tag{41}$$

[Robust function]

$$f_8^\pm = \lambda \sum_{p=1}^{P} \sum_{t=1}^{T} \sum_{h=1}^{H} P_{th} [\varepsilon_{pth}^\pm - \sum_{p=1}^{P} \sum_{h=1}^{H} P_{th} \cdot \varepsilon_{pth}^\pm + 2\theta_{pth}^\pm] \tag{42}$$

where,

$$\begin{aligned}
\varepsilon_{pth}^\pm &= PEC_{pt}^\pm \cdot DE_{pt}^\pm \cdot EF_{pt}^\pm + (PV_{pt}^\pm + PP_{pt}^\pm) \cdot DE_{pth}^\pm \\
&+ IE_t^\pm \cdot IPE_t^\pm + (YEH_{pth}^\pm \cdot A_{pt}^\pm + XEH_{pth}^\pm \cdot B_{pt}^\pm) \\
&+ \sum_{i=1}^{I} DE_{pth}^\pm \cdot \zeta_{ipt}^\pm \cdot \left(1 - \eta_{ipt}^\pm\right) \cdot DPC_{it}^\pm - DE_{pth}^\pm \cdot SU_{pt}^\pm
\end{aligned} \tag{43}$$

where $f^\pm$ is the objective of the proposed model (million yuan ¥); p is the power conversion technologies, p = 1, 2, 3, 4, and 5 for combined heat and power (CHP), hydroelectric power, solar photovoltaic power, wind power, and garbage power and biomass power, respectively; i denotes different atmospheric pollutants, i = 1, 2, 3, 4 for CO_2, SO_2, NOx, and particulate matter, respectively; t is the planning period; h denotes the electric-power demand level, h = 1 for low level, h = 2 for medium level, and h = 3 for high level, respectively. $Z_{pt}^\pm$ is the amount of energy resource consumption for power conversion technology p (PJ); $PEC_{pt}^\pm$ represents the energy price for technology p (million ¥/PJ); $PV_{pt}^\pm$ and $PP_{pt}^\pm$ are the variable cost for power generation and the penalty cost of excess power generation of technology p (million ¥/GWh); $AE_{pt}^\pm$ denotes the pre-regular electric-power generation by technology p (GWh); $DE_{pth}^\pm$ is the excess power generation by technology p under different electric-power deficiency levels h (GWh); $CV_{pt}^\pm$ represents the variable cost for heat generation by technology p (million ¥/PJ); $AH_{pt}^\pm$ is the amount of district heat supply by technology p (PJ); $DH_{pt}^\pm$ denotes the amount of district heat supply by expanded capacity $XEH_{pth}^\pm$ (PJ); $A_{pt}^\pm$ and $B_{pt}^\pm$ are the fixed-charge cost and variable cost for capacity expansion of technology p (million ¥); $SU_{pt}^\pm$ is the subsidy for new renewable energy generation p (million ¥/GW); $YEH_{pth}^\pm$ represents the binary variable for determining the capacity choice of technology p expansion (0 denotes no expansion; 1 represents expansion); $XEH_{pth}^\pm$ is the capacity expansion amount for technology p under different electric-power deficiency levels h (GW); $IE_{th}^\pm$ denotes imported power amount (GWh); $IPE_t^\pm$ is the cost of imported power (million ¥/GWh); $CPC_{it}^\pm$ and $DPC_{it}^\pm$ are the removal cost of pollutant i treatment and the penalty cost of excess pollutant i treatment for technology p (million¥/ton); $\zeta_{ipt}^\pm$ is the generation rate of pollutant from technology (ton/GWh).

Constraint:

[Constraints for electric-power supply and demand balance]

$$\sum_{p=1}^{P} \left(AE_{pt}^\pm + DE_{pth}^\pm\right) + IE_{th}^\pm \geq ADE_{th}^\pm, \forall t, h \tag{44}$$

$$(AE_{pt}^\pm + DE_{pth}^\pm) \leq ST_{pt}^\pm \cdot IC_{pt}^\pm, \forall p, t, h \tag{45}$$

$$AE_{pt}^\pm \geq DE_{pth}^\pm \geq 0, \forall p, t, h \tag{46}$$

$$IE_{th}^\pm \leq 40\% ADE_{th}^\pm, \forall t, h \tag{47}$$

[Constraints for the district heating supply and demand balance]

$$\sum_{p=1}^{P} (AH_{pt}^\pm + DH_{pt}^\pm) \geq TH_{th}^\pm, \forall t, h \tag{48}$$

$$AH_{pt}^\pm \geq DH_{pth}^\pm \geq 0, \forall p, t, h \tag{49}$$

[Constraint for combined heat and power generation balance]

$$Q^{\pm}m_{1t}^{\pm} = B_Q(\frac{AE_{1t}^{\pm}}{1 - ES^{\pm}}CE^{\pm} + \frac{AH_{1t}^{\pm}}{1 - HS^{\pm}}CH^{\pm}) \, \forall t; \tag{50}$$

[Constraints for the heat-to-electric ratio of cogeneration plant]

$$AH_{1t}^{\pm} + DH_{1t}^{\pm} = (AE_{1t}^{\pm} + DE_{1th}^{\pm}) \cdot \kappa^{\pm}, \forall t \tag{51}$$

$$XEH_{1th}^{\pm} \cdot ST_{1t}^{\pm} = DH_{1t}^{\pm} \cdot \kappa^{\pm}, \forall t \tag{52}$$

[Constraint for the total thermal efficiency of thermal power plant from national policy]

$$(AE_{1t}^{\pm} + AH_{1t}^{\pm}) \geq 45\% \cdot Q^{\pm}m_t^{\pm}, \forall t \tag{53}$$

[Constraints for environment capacity (CO2, PM, SO2, and NOx emission)]

$$\sum_{p=1}^{P} \left(AE_{pt}^{\pm} + DE_{pth}^{\pm}\right) \cdot \varsigma_{ipt}^{\pm} \cdot \left(1 - \eta_{ipt}^{\pm}\right) \leq MAGE_{it}, \, \forall i, t, h \tag{54}$$

[Constraints for installed capacity]

$$IC_{pt}^{\pm} = ICP_p + YEH_{pth}^{\pm} \cdot XEH_{pth}^{\pm} - CIC_{pt}^{\pm}, t = 1, \forall p, h \tag{55}$$

$$IC_{pt}^{\pm} = IC_{p(t-1)}^{\pm} + YEH_{pth}^{\pm} \cdot XEH_{pth}^{\pm} - CIC_{pt}^{\pm}, t > 1, \forall p, h \tag{56}$$

[Constraints for capacity expansion]

$$YEH_{pth}^{\pm} \begin{cases} = 1, \text{ if capacity expansion is undertaken} \\ = 0, \text{ otherwise} \end{cases}, \, \forall p, t, h \tag{57}$$

$$0 \leq XEH_{pth}^{\pm} \leq M_{pt} \cdot YEH_{pth}^{\pm}, \forall p, t, h \tag{58}$$

[Constraints for generation proportion of different technologies]

$$AE_{1t}^{\pm} + DE_{1th}^{\pm} \leq \gamma_t^{\pm} \cdot ADE_{dth}^{\pm}, \, \forall t, h \tag{59}$$

$$\sum_{p=3}^{5} \left(AE_{pt}^{\pm} + DE_{pth}^{\pm}\right) \geq \delta \cdot ADE_{dth}^{\pm}, \, \forall t, h \tag{60}$$

[Constraints for availabilities of energy resources]

$$\left(AE_{pt}^{\pm} + DE_{pth}^{\pm}\right) \cdot EE_{pt}^{\pm} \leq Z_{pt}^{\pm}, \, \forall p, t, h \tag{61}$$

[Robust constraints]

$$\varepsilon_{pth}^{\pm} - \sum_{p=1}^{P} \sum_{h=1}^{H} p_{th} \cdot \varepsilon_{pth}^{\pm} + \theta_{pth}^{\pm} \geq 0, \, \forall p, t, h \tag{62}$$

where $ADE_{th}^{\pm}$ denotes the electricity demand under different electric-power deficiency levels h during period t (GWh); $TH_{th}^{\pm}$ is the district heat demand under different deficiency levels h during period t (PJ); $Q^{\pm}$ is the heating value of coal (PJ/ton); $m_{1t}^{\pm}$ represents the coal quantity fed to combined heat and power (CHP) (ton); B_Q denotes the calorific value of coal (PJ/ton); $ES^{\pm}$ is the electricity consumption rate of thermal power plant; $CE^{\pm}$ is the standard coal consumption of power generation of thermal power plant (ton/PJ); $HS^{\pm}$ represents the heat loss of the facilities; $CH^{\pm}$ is the standard coal consumption of heat supply of thermal power plant (ton/PJ); $\kappa^{\pm}$ denotes heat-to-electric ratio;

$\gamma_t^\pm$ denotes the proportion of thermal power; $\eta_{ipt}^\pm$ is the removal efficiency of pollutant i from technology p; $\xi_{ipt}^\pm$ denotes the emission intensity of pollutant i from technology p (10^3 ton/GWh); $MAGE_{it}$ is the total allowable amount of pollutant i emission (10^3 ton); $M_{pt}^\pm$ and $N_{pt}^\pm$ are the constraints for the upper and lower capacity expansion bound of technology p (GW); $ST_{pt}^\pm$ is the operation hours of technology P in period t (h); δ denotes the percentage of power generation amount by renewable energy resources; ICP_p is the initial installed capacity of power conversion technology p (GW); $IC_{pt}^\pm$ represents the total installed capacity of technology p (GW); $CIC_{pt}^\pm$ denotes the closed installed capacity of "developing large units and suppressing small ones" in period t (GW); $EF_{pt}^\pm$ is the resources conversion efficiency of technology p (PJ/GWh).

The planning horizon is considered as being from 2016 to 2021, and divided into two periods with a 3-year interval for each period. The related technical-economic information was obtained through analyzing many representative energy-related governmental reports and plans. Table 1 presents power demands and the occurrence probabilities of each demand level (25%, 55%, and 20%). According to Zibo Statistics Bureau (from 1990 to 2014), and the forecasting information of electric-power demand by the government, three electricity generation targets are selected. Table 1 also shows the district heating demands during the planning horizon. To achieve the targets of renewable power generation and emission reduction, in the electric-power system, some scenarios are designed, which corresponds to environmental constraints and renewable power development constraints (i.e., renewable energy generation in period 1 and 2 accounts for 5% and 10% of the total regional power consumption, respectively).

Table 1. Regional electricity and heat demand during the planning period.

Energy Demand	Demand Level	Probability (%)	T = 1	T = 2
Electricity demand (10^3 GWh)	Low	20	[97.11, 98.53]	[97.73, 99.11]
	Medium	60	[98.53, 100.14]	[99.21, 100.60]
	High	20	[99.90, 100.60]	[101.00, 102.60]
District heat quantity (PJ)	Low	20	[253.59, 259.59]	[255.00, 263.00]
	Medium	60	[278.68, 288.68]	[285.00, 293.00]
	High	20	[288.87, 297.87]	[295.00, 302.00]

4. Result Analysis and Discussion

4.1. Electricity-Generation Plan

Tables 2–5 present the optimal solutions of electric-power generation schemes of different technologies with different λ values, under different demand levels, during the whole planning horizon. The optimal combined electricity and heat generation targets in period 2 would be greater than that in period 1. In period 1, the generation amount of combined electricity and heat would be 56.99×10^3 GWh in period 1, and 59.98×10^3 GWh in period 2 under different λ values. Furthermore, power generation amount of CHP would increase. For example, in period 1, under medium demand level, power generation amount by CHP would be 58.23×10^3 GWh, 56.99×10^3 GWh, 57.22×10^3 GWh, and 56.99×10^3 GWh, as λ is fixed with the values of 0, 1, 5 and 50, respectively; under medium–medium level (with the probability of 30.25%) in period 2, power generation amount would be $(68.28, 68.72) \times 10^3$ GWh, 61.51×10^3 GWh, 59.98×10^3 GWh, and 61.22×10^3 GWh, respectively. It indicated that the CHP is a more economical and stable way for power supply with the demand of electricity increasing, and along with regional electric-power structure optimization, the combined heat and power would still be the main choice for supporting regional electric-power supply.

Table 2. The optimized power generation schemes under λ = 0.

Technology	Level	Probability (%)	Optimized Generation Target (GWh)	Optimized Shortage Quantity (GWh)	Optimized Generation Quantity (GWh)
CHP	L	25	56,989.23	1240.12	58,229.35
	M	55	56,989.23	1240.12	58,229.35
	H	20	56,989.23	1240.12	58,229.35
	L-L	6.25	59,978.03	[8302.54, 8737.05]	[68,280.57, 68,715.08]
	L-M	13.75	59,978.03	[8302.54, 8737.05]	[68,280.57, 68,715.08]
	L-H	5	59,978.03	[8302.54, 8737.05]	[68,280.57, 68,715.08]
	M-L	13.75	59,978.03	[8302.54, 8737.05]	[68,280.57, 68,715.08]
	M-M	30.25	59,978.03	[8302.54, 8737.05]	[68,280.57, 68,715.08]
	M-H	11	59,978.03	[8302.54, 8737.05]	[68,280.57, 68,715.08]
	H-L	5	59,978.03	[8302.54, 8737.05]	[68,280.57, 68,715.08]
	H-M	11	59,978.03	[8302.54, 8737.05]	[68,280.57, 68,715.08]
	H-H	4	59,978.03	[8302.54, 8737.05]	[68,280.57, 68,715.08]
Hydropower	L	25	29.01	29.01	58.02
	M	55	29.01	29.01	58.02
	H	20	29.01	29.01	58.02
	L-L	6.25	30.85	30.85	61.70
	L-M	13.75	30.85	30.85	61.70
	L-H	5	30.85	30.85	61.70
	M-L	13.75	30.85	30.85	61.70
	M-M	30.25	30.85	30.85	61.70
	M-H	11	30.85	30.85	61.70
	H-L	5	30.85	30.85	61.70
	H-M	11	30.85	30.85	61.70
	H-H	4	30.85	30.85	61.70
Solar power	L	25	303.74	303.74	607.48
	M	55	303.74	303.74	607.48
	H	20	303.74	303.74	607.48
	L-L	6.25	406.53	[279.34, 315.44]	[685.87, 721.97]
	L-M	13.75	406.53	[279.34, 315.44]	[685.87, 721.97]
	L-H	5	406.53	[279.34, 315.44]	[685.87, 721.97]
	M-L	13.75	406.53	[279.34, 315.44]	[685.87, 721.97]
	M-M	30.25	406.53	[279.34, 315.44]	[685.87, 721.97]
	M-H	11	406.53	[279.34, 315.44]	[685.87, 721.97]
	H-L	5	406.53	[279.34, 315.44]	[685.87, 721.97]
	H-M	11	406.53	[279.34, 315.44]	[685.87, 721.97]
	H-H	4	406.53	[279.34, 315.44]	[685.87, 721.97]
Wind power	L	25	1691.76	1691.76	3383.52
	M	55	1691.76	1691.76	3383.52
	H	20	1691.76	1691.76	3383.52
	L-L	6.25	2077.83	2077.83	4155.66
	L-M	13.75	2077.83	2077.83	4155.66
	L-H	5	2077.83	2077.83	4155.66
	M-L	13.75	2077.83	2077.83	4155.66
	M-M	30.25	2077.83	2077.83	4155.66
	M-H	11	2077.83	2077.83	4155.66
	H-L	5	2077.83	2077.83	4155.66
	H-M	11	2077.83	2077.83	4155.66
	H-H	4	2077.83	2077.83	4155.66

Table 2. *Cont.*

Technology	Level	Probability (%)	Optimized Generation Target (GWh)	Optimized Shortage Quantity (GWh)	Optimized Generation Quantity (GWh)
	L	25	877.53	877.53	1755.06
	M	55	877.53	877.53	1755.06
	H	20	877.53	877.53	1755.06
	L-L	6.25	2522.63	2522.63	5045.26
Biomass and	L-M	13.75	2522.63	2522.63	5045.26
garbage power	L-H	5	2522.63	2522.63	5045.26
	M-L	13.75	2522.63	2522.63	5045.26
	M-M	30.25	2522.63	2522.63	5045.26
	M-H	11	2522.63	2522.63	5045.26
	H-L	5	2522.63	2522.63	5045.26
	H-M	11	2522.63	2522.63	5045.26
	H-H	4	2522.63	2522.63	5045.26

Table 3. The optimized power generation schemes under $\lambda = 1$.

Technology	Level	Probability (%)	Optimized Generation Target (GWh)	Optimized Shortage Quantity (GWh)	Optimized Generation Quantity (GWh)
	L	25	56,989.23	0	56,989.23
	M	55	56,989.23	0	56,989.23
	H	20	56,989.23	0	56,989.23
	L-L	6.25	59,978.03	1536.75	61,514.78
	L-M	13.75	59,978.03	1536.75	61,514.78
CHP	L-H	5	59,978.03	1536.75	61,514.78
	M-L	13.75	59,978.03	1536.75	61,514.78
	M-M	30.25	59,978.03	1536.75	61,514.78
	M-H	11	59,978.03	1536.75	61,514.78
	H-L	5	59,978.03	1536.75	61,514.78
	H-M	11	59,978.03	1536.75	61,514.78
	H-H	4	59,978.03	1536.75	61,514.78
	L	25	29.01	29.01	58.02
	M	55	29.01	29.01	58.02
	H	20	29.01	29.01	58.02
	L-L	6.25	30.85	30.85	61.70
	L-M	13.75	30.85	30.85	61.70
Hydropower	L-H	5	30.85	30.85	61.70
	M-L	13.75	30.85	30.85	61.70
	M-M	30.25	30.85	30.85	61.70
	M-H	11	30.85	30.85	61.70
	H-L	5	30.85	30.85	61.70
	H-M	11	30.85	30.85	61.70
	H-H	4	30.85	30.85	61.70
	L	25	303.74	303.74	607.48
	M	55	303.74	303.74	607.48
	H	20	303.74	303.74	607.48
	L-L	6.25	406.53	[255.63, 290.48]	[662.16, 697.01]
Solar power	L-M	13.75	406.53	[255.63, 290.48]	[662.16, 697.01]
	L-H	5	406.53	[255.63, 290.48]	[662.16, 697.01]
	M-L	13.75	406.53	[255.63, 290.48]	[662.16, 697.01]
	M-M	30.25	406.53	[255.63, 290.48]	[662.16, 697.01]

Table 3. *Cont.*

Technology	Level	Probability (%)	Optimized Generation Target (GWh)	Optimized Shortage Quantity (GWh)	Optimized Generation Quantity (GWh)
	M-H	11	406.53	[255.63, 290.48]	[662.16, 697.01]
	H-L	5	406.53	[255.63, 290.48]	[662.16, 697.01]
	H-M	11	406.53	[255.63, 290.48]	[662.16, 697.01]
	H-H	4	406.53	[255.63, 290.48]	[662.16, 697.01]
	L	25	1691.76	1691.76	3383.52
	M	55	1691.76	1691.76	3383.52
	H	20	1691.76	1691.76	3383.52
	L-L	6.25	2077.83	2077.83	4155.66
	L-M	13.75	2077.83	2077.83	4155.66
Wind power	L-H	5	2077.83	2077.83	4155.66
	M-L	13.75	2077.83	2077.83	4155.66
	M-M	30.25	2077.83	2077.83	4155.66
	M-H	11	2077.83	2077.83	4155.66
	H-L	5	2077.83	2077.83	4155.66
	H-M	11	2077.83	2077.83	4155.66
	H-H	4	2077.83	2077.83	4155.66
	L	25	877.53	0	877.53
	M	55	877.53	[0, 80.6]	[877.53, 958.13]
	H	20	877.53	[68.45, 103.45]	[945.98, 980.98]
	L-L	6.25	2534.49	2212.79	4747.28
Biomass and	L-M	13.75	2534.49	2212.79	4747.28
garbage power	L-H	5	2534.49	2212.79	4747.28
	M-L	13.75	2534.49	2357.88	4892.37
	M-M	30.25	2534.49	2357.88	4892.37
	M-H	11	2534.49	2357.88	4892.37
	H-L	5	2534.49	2534.49	5068.98
	H-M	11	2534.49	2534.49	5068.98
	H-H	4	2534.49	2534.49	5068.98

Table 4. The optimized power generation schemes under $\lambda = 5$.

Technology	Level	Probability (%)	Optimized Generation Target (GWh)	Optimized Shortage Quantity (GWh)	Optimized Generation Quantity (GWh)
	L	25	56,989.23	226.47	57,215.70
	M	55	56,989.23	226.47	57,215.70
	H	20	56,989.23	226.47	57,215.70
	L-L	6.25	59,978.03	0	59,978.03
	L-M	13.75	59,978.03	0	59,978.03
CHP	L-H	5	59,978.03	0	59,978.03
	M-L	13.75	59,978.03	0	59,978.03
	M-M	30.25	59,978.03	0	59,978.03
	M-H	11	59,978.03	0	59,978.03
	H-L	5	59,978.03	0	59,978.03
	H-M	11	59,978.03	0	59,978.03
	H-H	4	59,978.03	0	59,978.03

Table 4. *Cont.*

Technology	Level	Probability (%)	Optimized Generation Target (GWh)	Optimized Shortage Quantity (GWh)	Optimized Generation Quantity (GWh)
Hydropower	L	25	29.01	29.01	58.02
	M	55	29.01	29.01	58.02
	H	20	29.01	29.01	58.02
	L-L	6.25	30.85	30.85	61.70
	L-M	13.75	30.85	30.85	61.70
	L-H	5	30.85	30.85	61.70
	M-L	13.75	30.85	30.85	61.70
	M-M	30.25	30.85	30.85	61.70
	M-H	11	30.85	30.85	61.70
	H-L	5	30.85	30.85	61.70
	H-M	11	30.85	30.85	61.70
	H-H	4	30.85	30.85	61.70
Solar power	L	25	303.74	303.74	607.48
	M	55	303.74	303.74	607.48
	H	20	303.74	303.74	607.48
	L-L	6.25	406.53	[255.63, 290.48]	[662.16, 697.01]
	L-M	13.75	406.53	[255.63, 290.48]	[662.16, 697.01]
	L-H	5	406.53	[255.63, 290.48]	[662.16, 697.01]
	M-L	13.75	406.53	[255.63, 290.48]	[662.16, 697.01]
	M-M	30.25	406.53	[255.63, 290.48]	[662.16, 697.01]
	M-H	11	406.53	[255.63, 290.48]	[662.16, 697.01]
	H-L	5	406.53	[255.63, 290.48]	[662.16, 697.01]
	H-M	11	406.53	[255.63, 290.48]	[662.16, 697.01]
	H-H	4	406.53	[255.63, 290.48]	[662.16, 697.01]
Wind power	L	25	1691.76	1691.76	3383.52
	M	55	1691.76	1691.76	3383.52
	H	20	1691.76	1691.76	3383.52
	L-L	6.25	2077.83	2077.83	4155.66
	L-M	13.75	2077.83	2077.83	4155.66
	L-H	5	2077.83	2077.83	4155.66
	M-L	13.75	2077.83	2077.83	4155.66
	M-M	30.25	2077.83	2077.83	4155.66
	M-H	11	2077.83	2077.83	4155.66
	H-L	5	2077.83	2077.83	4155.66
	H-M	11	2077.83	2077.83	4155.66
	H-H	4	2077.83	2077.83	4155.66
Biomass and garbage power	L	25	945.98	0	945.98
	M	55	945.98	[0, 12.15]	[945.98, 958.13]
	H	20	945.98	[0, 35]	[945.98, 980.98]
	L-L	6.25	2787.65	1959.62	4747.27
	L-M	13.75	2787.65	1959.62	4747.27
	L-H	5	2787.65	1959.62	4747.27
	M-L	13.75	2787.65	2104.72	4892.37
	M-M	30.25	2787.65	2104.72	4892.37
	M-H	11	2787.65	2104.72	4892.37
	H-L	5	2787.65	2281.33	5068.98
	H-M	11	2787.65	2281.33	5068.98
	H-H	4	2787.65	2281.33	5068.98

Table 5. The optimized power generation schemes under λ = 50.

Technology	Level	Probability (%)	Optimized Generation Target (GWh)	Optimized Shortage Quantity (GWh)	Optimized Generation Quantity (GWh)
	L	25	56,989.23	0	56,989.23
	M	55	56,989.23	0	56,989.23
	H	20	56,989.23	0	56,989.23
	L-L	6.25	59,978.03	1244.54	61,222.57
	L-M	13.75	59,978.03	1244.54	61,222.57
CHP	L-H	5	59,978.03	1244.54	61,222.57
	M-L	13.75	59,978.03	1244.54	61,222.57
	M-M	30.25	59,978.03	1244.54	61,222.57
	M-H	11	59,978.03	1244.54	61,222.57
	H-L	5	59,978.03	1244.54	61,222.57
	H-M	11	59,978.03	1244.54	61,222.57
	H-H	4	59,978.03	1244.54	61,222.57
	L	25	29.01	29.01	58.02
	M	55	29.01	29.01	58.02
	H	20	29.01	29.01	58.02
	L-L	6.25	30.85	30.85	61.70
	L-M	13.75	30.85	30.85	61.70
Hydropower	L-H	5	30.85	30.85	61.70
	M-L	13.75	30.85	30.85	61.70
	M-M	30.25	30.85	30.85	61.70
	M-H	11	30.85	30.85	61.70
	H-L	5	30.85	30.85	61.70
	H-M	11	30.85	30.85	61.70
	H-H	4	30.85	30.85	61.70
	L	25	303.74	303.74	607.48
	M	55	303.74	303.74	607.48
	H	20	303.74	303.74	607.48
	L-L	6.25	406.53	[255.63, 272.34]	[662.16, 678.87]
	L-M	13.75	406.53	[255.63, 272.34]	[662.16, 678.87]
Solar power	L-H	5	406.53	[255.63, 272.34]	[662.16, 678.87]
	M-L	13.75	406.53	[255.63, 272.12]	[662.16, 678.65]
	M-M	30.25	406.53	[255.63, 272.12]	[662.16, 678.65]
	M-H	11	406.53	[255.63, 272.12]	[662.16, 678.65]
	H-L	5	406.53	[255.63, 290.11]	[662.16, 696.64]
	H-M	11	406.53	[255.63, 290.11]	[662.16, 696.64]
	H-H	4	406.53	[255.63, 290.11]	[662.16, 696.64]
	L	25	1691.76	1691.76	3383.52
	M	55	1691.76	1691.76	3383.52
	H	20	1691.76	1691.76	3383.52
	L-L	6.25	2077.83	2077.83	4155.66
	L-M	13.75	2077.83	2077.83	4155.66
Wind power	L-H	5	2077.83	2077.83	4155.66
	M-L	13.75	2077.83	2077.83	4155.66
	M-M	30.25	2077.83	2077.83	4155.66
	M-H	11	2077.83	2077.83	4155.66
	H-L	5	2077.83	2077.83	4155.66
	H-M	11	2077.83	2077.83	4155.66
	H-H	4	2077.83	2077.83	4155.66

Table 5. *Cont.*

Technology	Level	Probability (%)	Optimized Generation Target (GWh)	Optimized Shortage Quantity (GWh)	Optimized Generation Quantity (GWh)
	L	25	945.98	0	945.98
	M	55	945.98	[0, 12.15]	[945.978, 958.13]
	H	20	945.98	[0, 35]	[945.978, 980.98]
	L-L	6.25	2787.65	1959.62	4747.27
	L-M	13.75	2787.65	1959.62	4747.27
Biomass and	L-H	5	2787.65	1959.62	4747.27
garbage power	M-L	13.75	2787.65	2104.72	4892.37
	M-M	30.25	2787.65	2104.72	4892.37
	M-H	11	2787.65	2104.72	4892.37
	H-L	5	2787.65	2281.33	5068.98
	H-M	11	2787.65	2281.33	5068.98
	H-H	4	2787.65	2281.33	5068.98

Among these renewable power generation technologies, clean electricity would mainly come from solar power, wind power, and biomass and garbage power (BGP). The optimized electricity generation for wind power would be 3.38×10^3 GWh and 4.16×10^3 GWh in periods 1 and 2, respectively. The wind power would play a significant role in renewable power development during the planning horizon. For example, in period 1, wind power generation would occupy about 3% of total electricity consumption, and 60% of total renewable power generation under different demand level; in period 2, the proportion would increase from about 3% to 4% of total electricity consumption, and be 40% of total renewable power generation. Since wind power possesses the characteristic of cleanliness and the condition of convenience in this region, wind power would be developed as a priority. In addition, BGP power generation would increase significantly during the whole planning horizon. For example, in period 1 under medium level (with the probability of 55%), power generation amount of BGP would be 1.76×10^3 GWh, (877.53, 958.13) GWh, (945.98, 958.13) GWh, and (945.98, 958.13) GWh under λ with the values of 0, 1, 5, and 50, respectively; in period 2 under medium–medium level, power generation amount of BGP would be 5.05×10^3 GWh, 5.07×10^3 GWh, 5.07×10^3 GWh, and 5.07×10^3 GWh with λ fixed as 0, 1, 5 and 50, respectively. The proportion of biomass and garbage power generation would rise from about 1% in period 1, to 5% in period 2 of total electricity consumption, and 20% in period 1 to 50% in period 2 of total renewable power generation. In Zibo city, the hydropower would have a smaller scale under water resource and geography limitation. In general, renewable power generation amount would change as λ values vary, and the stability of the regional electric-power supply would be enhanced as the total renewable power generation amount increases.

Figures 2–5 show the optimized solutions for electric-power generation schemes under different λ values. Electric-power generation amount of CHP would be decreased as λ increases. For example, in period 2 under medium–medium level, electric-power amount generated by CHP would be $(68.28, 68.72) \times 10^3$ GWh, 61.51×10^3 GWh, 59.98×10^3 GWh, and 61.22×10^3 GWh under λ fixed as 0, 1, 5, and 50, respectively. It indicated that the risk of system failure, which means higher CO_2 and pollutants discharged from cogeneration exceeding the regulated limitation, would decrease as λ increases. In general, relatively lower power generation of CHP would promote emissions reduction and evade the risk of regional energy system.

As shown in Figure 3, solar power generation amount would be decreased as λ value increases. For example, in period 2 under medium–medium level, the electricity generated by solar power would be (685.87, 721.97) GWh, (662.16, 697.01) GWh, (662.16, 697.01) GWh, and (662.16, 678.65) GWh as λ is fixed with the values of 0, 1, 5, and 50, respectively. Since the regional power supply of solar power has the characteristic of instability and higher cost, the stability and security of system power supply would increase as λ increases. Electric power generated by BGP would decrease as λ increases (Figure 5). For instance, under λ fixed with the values of 0, 1, 5, and 50, power generation amount of BGP would be 1.76×10^3 GWh, (877.53, 958.13) GWh, (945.98, 958.13) GWh, and (945.98, 958.13) GWh in period 1 under medium level, respectively; the generation amount would be 5.05×10^3 GWh, 5.07×10^3 GWh, 5.07×10^3 GWh, and 5.07×10^3 GWh under medium–medium level in period 2, respectively. A higher power generation of BGP would lead to a higher pollutants and CO_2 emission, which would violate environmental constraints of the system. As λ increases, the power generation of BGP would be reduced. In summary, the total renewable power generation amount would increase as λ values increase. Thus, as λ values increases, the system failure risk would be lessened; meanwhile, the security and stability would be enhanced.

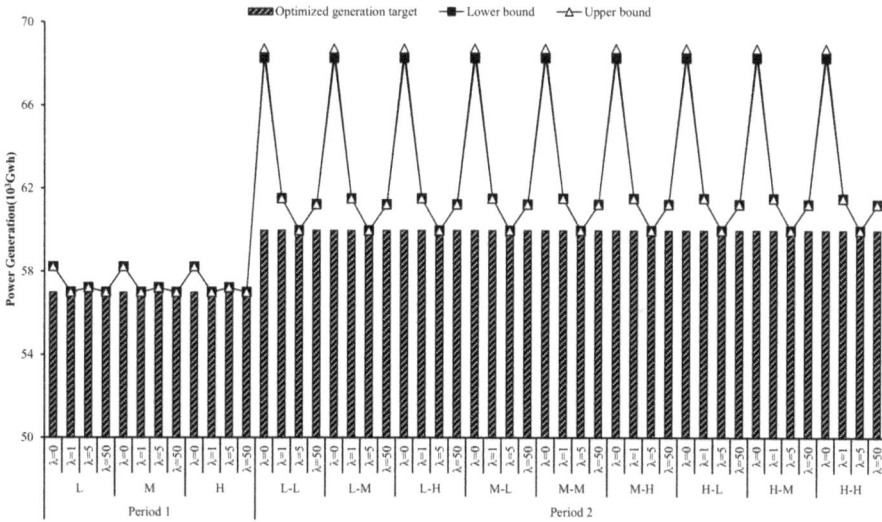

Figure 2. The optimized cogeneration operation schemes during the planning horizon.

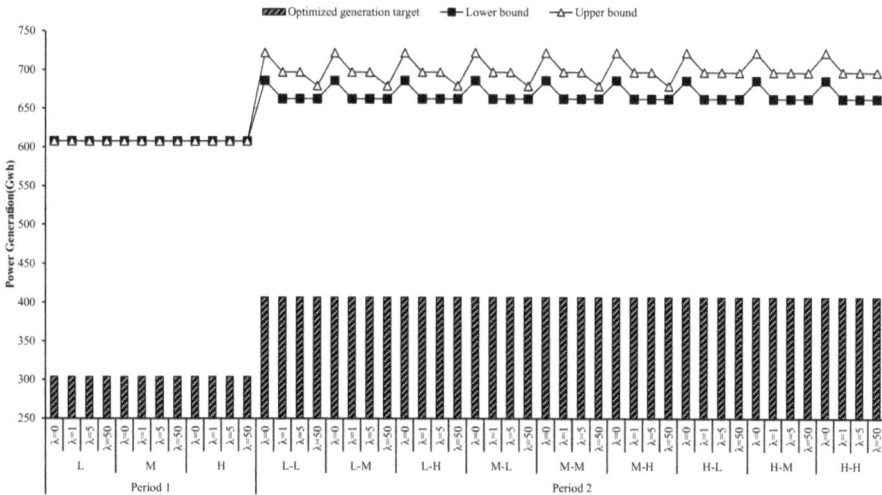

Figure 3. The optimized solar power generation amount in planning periods.

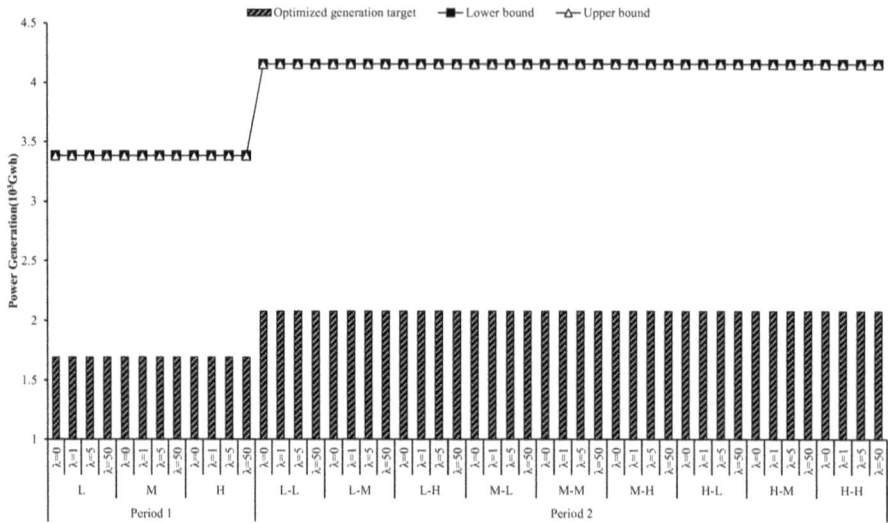

Figure 4. The optimized wind power generation during the whole planning horizon.

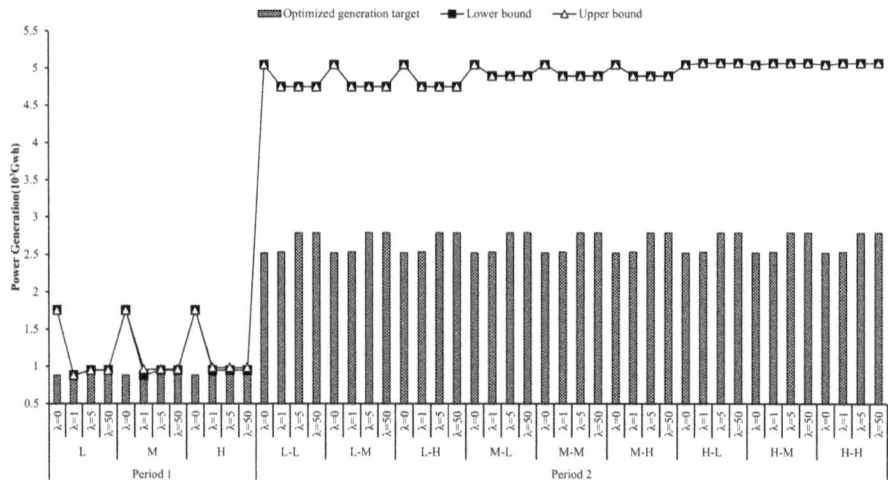

Figure 5. Optimized biomass power generation amount in the planning periods.

4.2. Imported Electricity Scheme

Figure 6 presents the imported electric power amount during the planning horizon. It would decrease from period 1 to 2 under different power demand levels. For example, with λ fixed as 0,1,5, and 50, the imported power amount would be $(34.5, 35.41) \times 10^3$ GWh, $(36.62, 38.15) \times 10^3$ GWh, $(36.32, 37.92) \times 10^3$ GWh, and $(36.55, 38.15) \times 10^3$ GW h under medium level in period 1, respectively; $(20.98, 21.9) \times 10^3$ GWh, $(28.1, 29.46) \times 10^3$ GWh, $(29.46, 30.82) \times 10^3$ GWh, and $(28.21, 29.59) \times 10^3$ GWh under medium–medium demand level in period 2, respectively. It indicated that the imported power amount would be decreased with regional power generation and power structure adjustment increasing, and the amount of imported electricity would rise as λ increases. For instance, under medium level in period 1, the amount would be

$(34.5, 35.41) \times 10^3$ GWh under $\lambda = 0$ and $(36.55, 38.15) \times 10^3$ GWh under $\lambda = 50$. As a result, it would lead to a smaller system risk and enhanced system feasibility, which could also promote the energy conservation and emissions reduction to some degree.

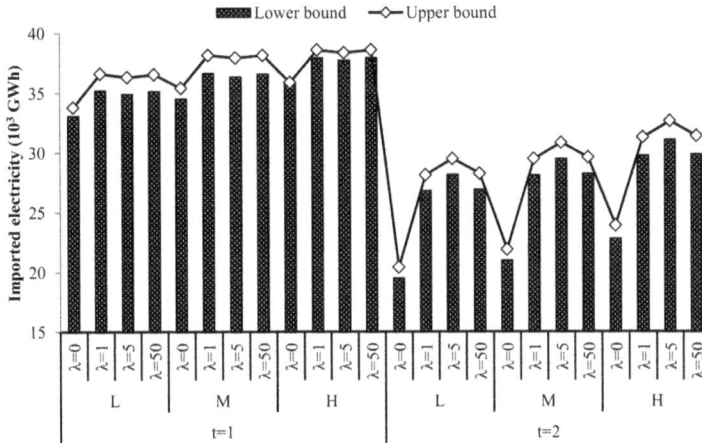

Figure 6. The imported electricity amount under different λ values.

4.3. CO₂ and Air Pollution Control

Table 6 shows the solutions of optimized air pollutants and CO_2 emission. The air pollutants and CO_2 emission amount would decrease. For example, under $\lambda = 5$, the amount of CO_2 emissions would decrease from $(60.54, 61.01) \times 10^6$ ton in period 1 to $(55.6, 56.17) \times 10^6$ ton in period 2; the amount of SO_2 emissions would decrease from $(69.03, 80.23) \times 10^3$ ton in period 1 to $(45.35, 66.78) \times 10^3$ ton in period 2; the amount of NO_x emissions would be $(59.27, 83.71) \times 10^3$ ton and $(35.61, 70.75) \times 10^3$ ton in period 1 and 2; the amount of PM_{10} emissions would decrease from $(9.56, 12.43) \times 10^3$ ton in period 1 to $(4.89, 7.02) \times 10^3$ ton in period 2, respectively. The reasons for decreasing emission are that firstly, the technology and facilities would be updated to reduce the average emissions level; secondly, due to the power structure optimization in the first period, renewable energy has been developed to some degree, which could make contributions to energy conservation and emissions reduction. In addition, the effect of emission mitigation could be better under considering system risk aversion. It indicates that the results would lead to a lower system risk and more robust regional energy system, which is important for achieving sustainable development and better environment quality.

Table 6. The amount of CO_2 and air pollution emissions under different λ values.

Gaseous Emission	λ Level	T = 1	T = 2
CO_2 (10^6 ton)	$\lambda = 0$	[61.88, 63.11]	[63.07, 64.10]
	$\lambda = 1$	[60.3, 60.77]	[56.82, 57.40]
	$\Lambda = 5$	[60.54, 61.01]	[55.60, 56.17]
	$\lambda = 50$	[60.31, 60.77]	[56.72, 57.30]
SO_2 (10^3 ton)	$\lambda = 0$	[70.35, 82.75]	[51.58, 76.43]
	$\lambda = 1$	[68.76, 79.91]	[46.37, 68.28]
	$\lambda = 5$	[69.03, 80.23]	[45.35, 66.78]
	$\lambda = 50$	[68.76, 79.91]	[46.28, 68.16]

Table 6. *Cont.*

Gaseous Emission	λ Level	T = 1	T = 2
NO$_x$ (10^3 ton)	λ = 0	[60.36, 86.29]	[40.51, 81.00]
	λ = 1	[59.04, 83.38]	[36.41, 72.34]
	λ = 5	[59.27, 83.71]	[35.61, 70.75]
	λ = 50	[59.03, 83.38]	[36.35, 72.21]
PM (10^3 ton)	λ = 0	[9.74, 12.82]	[5.57, 8.04]
	λ = 1	[9.53, 12.38]	[5.00, 7.18]
	λ = 5	[9.56, 12.43]	[4.89, 7.02]
	λ = 50	[9.52, 12.38]	[5.00, 7.17]

4.4. System Cost

Figure 7 shows the total system costs under different scenarios during the planning periods. The energy system cost in Zibo city would have a slight increase trend as λ levels increase. For instance, under the scenarios of λ with the values of 0, 1, 5, and 50, the system cost would be RMB¥ (490.63, 651.16) × 10^9, RMB¥ (499.65, 659.71) × 10^9, RMB¥ (502.62, 662.94) × 10^9, and RMB¥ (502.82, 663.56) × 10^9, respectively. As λ levels increasing, the system failure risk would be reduced, and the system cost would be increased. Conversely, a lower λ level would bring about a higher system risk and a lower system cost. It indicated that if the decision-makers aim to lower costs, a higher system risk may occur.

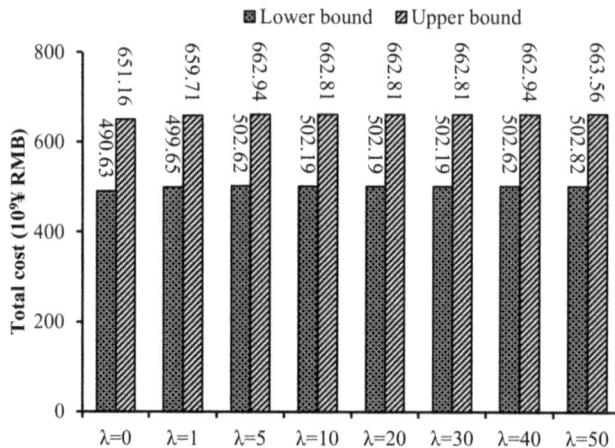

Figure 7. Net system cost under different scenarios.

Based on the above analyses, these indicated that the optimized solutions are able to support regional energy system management for making integrated schemes of power generation, capacity expansion, air pollutant and CO$_2$ emission reduction under different renewable energy development targets and environmental quality requirements. The solutions with lower and upper bounds are helpful for generating decision alternatives representing various options. Cost–risk analysis can be obtained through integrated the stochastic robust optimization method into the multistage stochastic programming in regional energy system management.

5. Conclusions

In this study, a multistage stochastic inexact robust programming was proposed for supporting regional electric-power system structure optimization and management. The model covered the district heating supply, power generation, and air pollutant mitigation coupled with relevant technique constraints and governmental policies. Comparing the solutions optimized by the model with the strategies carried out in the real world, the former emerged with obvious advantages, which can lead to a more prosperous future. In addition, the developed method could be valuable for obtaining trade-off schemes between system economy and risks that are introduced by system's uncertainties according to decision-makers' willingness. An energy system structure management of Zibo City, China, is used as a case study for verifying the efficiency of the developed model. Optimized schemes of power generation, capacity expansion, air pollutant and CO_2 emission reduction, and system cost were analyzed. The results indicated that under different requirements of renewable energy development, and pollutant and CO_2 mitigation, traditional power generation technology would still be increased, attributing to its lower costs and traditional energy resources structure based on the thermal power generation. In addition, renewable energy would also play an important role in solving energy, resource, and environmental pressures; renewable power generation amount would be rising continuously, though it might develop slowly for a certain period of time.

However, a number of limitations also exist in the proposed model of this study. Firstly, in the optimization model, many energy industrial processes are not considered, and only generation processes and energy-related environmental problems are involved in this study. In order to obtain more comprehensive management schemes, more energy development and utilization patterns could be considered. Second, compared with other optimization methods, the model would be infeasible in addressing the high uncertainties in the model parameters; and through introducing different λ values in the model, regional energy-managers cannot directly obtain suitable management schemes. Therefore, further research can strengthen knowledge and mitigate these limitations in the future.

Author Contributions: Y.X., L.W. and L.J. designed the manuscript and developed the models; Y.X. drafted the manuscript; L.W. and D.X. collected the data and revised the manuscript; G.H., Y.X. and D.X. checked the content and revised the manuscript. All authors made contributions to the study and the writing of the manuscript.

Funding: The National Natural Science Foundation of China (51609003 and 71603016), the China Postdoctoral Science Foundation funded project (Grand No.2015M580046 and Grand No.2015M580034), and the Fundamental Research Funds for the Central Universities (FRF-BD-18-015A).

Acknowledgments: This research was supported by the National Natural Science Foundation of China (51609003 and 71603016), the China Postdoctoral Science Foundation funded project (Grand No.2015M580046 and Grand No.2015M580034), and the Fundamental Research Funds for the Central Universities (FRF-BD-18-015A). The authors are grateful to the anonymous reviewers and editors for their valuable comments and suggestions.

Conflicts of Interest: The authors declare that there are no conflicts of interest regarding the publication of this article.

References

1. Cristóbal, J.; Guillén-Gosálbez, G.; Jiménez, L. Multi-objective optimization of coal-fired electricity production with CO_2 capture. *Appl. Energy* **2012**, *98*, 266–272. [CrossRef]
2. Dubreuil, A.; Assoumou, E.; Bouckaert, S.; Selosse, S.; MaZi, N. Water modelling in an energy optimization framework—The water-scarce middle east context. *Appl. Energy* **2013**, *101*, 268–279. [CrossRef]
3. Cheng, G.H.; Huang, G.H.; Dong, C.; Baetz, B.W.; Li, Y.P. Interval recourse linear programming for resources and environmental systems management under uncertainty. *J. Environ. Inform.* **2017**, *30*, 119–136. [CrossRef]
4. Luhandjulaa, M.K. Fuzzy stochastic linear programming: Survey and future research directions. *Eur. J. Oper. Res.* **2006**, *174*, 1353–1367. [CrossRef]
5. Mackay, R.M.; Probert, S.D. Energy and environmental policies of the developed and developing countries within the evolving Oceania and South-East Asia trading bloc. *Appl. Energy* **1995**, *51*, 369–400. [CrossRef]
6. Price, T.J.; Probert, S.D. Taiwan's energy and environmental policies: Past, present and future. *Appl. Energy* **1995**, *50*, 41–68. [CrossRef]

7. Shimazaki, Y.; Akisawa, A.; Kashiwagi, T. A model analysis of clean development mechanisms to reduce both CO2 and SO2 emissions between Japan and China. *Appl. Energy* **2011**, *66*, 311–324. [CrossRef]
8. Chaaban, F.B.; Mezher, T.; Ouwayjan, M. Options for emissions reduction from power plants: An economic evaluation. *Int. J. Electr. Power Energy Syst.* **2004**, *26*, 57–63. [CrossRef]
9. Jebaraj, S.; Iniyan, S. A review of energy models. *Renew. Sustain. Energy Rev.* **2006**, *10*, 281–311. [CrossRef]
10. Cai, Y.P.; Tan, Q.; Huang, G.H.; Yang, Z.F.; Lin, Q.G. Community-scale renewable energy systems planning under uncertainty—An interval chance-constrained programming approach. *Renew. Sustain. Energy Rev.* **2009**, *13*, 721–735. [CrossRef]
11. Dincer, I.; Rosen, M.A. Exergy, energy, environment and sustainable development. *Appl. Energy* **2014**, *64*, 427–440. [CrossRef]
12. Mukherjee, U.; Maroufmashat, A.; Narayan, A.; Elkamel, A.; Fowler, M. A stochastic programming approach for the planning and operation of a power to gas energy hub with multiple energy recovery pathways. *Energies* **2017**, *10*, 868. [CrossRef]
13. Li, W.; Bao, Z.; Huang, G.H.; Xie, Y.L. An inexact credibility chance-constrained integer programming for greenhouse gas mitigation management in regional electric power system under uncertainty. *J. Environ. Inform.* **2018**, *31*, 111–122. [CrossRef]
14. Tang, Z.C.; Xia, Y.J.; Xue, Q.; Liu, J. A non-probabilistic solution for uncertainty and sensitivity analysis on techno-economic assessments of biodiesel production with interval uncertainties. *Energies* **2018**, *11*, 588. [CrossRef]
15. Cai, Y.P.; Huang, G.H.; Lu, H.W.; Yang, Z.F.; Tan, Q. I-VFRP: An interval-valued fuzzy robust programming approach for municipal waste-management planning under uncertainty. *Eng. Optim.* **2009**, *41*, 399–418. [CrossRef]
16. Li, Y.F.; Li, Y.P.; Huang, G.H.; Chen, X. Energy and environmental systems planning under uncertainty—An inexact fuzzy-stochastic programming approach. *Appl. Energy* **2010**, *87*, 3189–3211. [CrossRef]
17. Li, Y.P.; Huang, G.H.; Chen, X. Planning regional energy system in association with greenhouse gas mitigation under uncertainty. *Appl. Energy* **2011**, *88*, 599–611. [CrossRef]
18. Huang, C.Z.; Nie, S.; Guo, L.; Fan, Y.R. Inexact fuzzy stochastic chance constraint programming for emergency evacuation in Qinshan nuclear power plant under uncertainty. *J. Environ. Inform.* **2017**, *30*, 63–78. [CrossRef]
19. Sheikhahmadi, P.; Mafakheri, R.; Bahramara, S.; Damavandi, M.Y.; Catalão, J.P.S. Risk-based two-stage stochastic optimization problem of micro-grid operation with renewable and incentive-based demand response programs. *Energies* **2018**, *11*, 610. [CrossRef]
20. Li, Y.P.; Huang, G.H.; Nie, S.L. An interval-parameter multi-stage stochastic programming model for water resources management under uncertainty. *Adv. Water Resour.* **2006**, *29*, 776–789. [CrossRef]
21. Li, G.C.; Huang, G.H.; Lin, Q.G.; Chen, Y.M.; Zhang, X.D. Development of an interval multi-stage stochastic programming model for regional energy systems planning and GHG emission control under uncertainty. *Int. J. Energy Res.* **2012**, *36*, 1161–1174. [CrossRef]
22. Hu, Q.; Huang, G.H.; Cai, Y.P.; Xu, Y. Energy and environmental systems planning with recourse: Inexact stochastic programming model containing fuzzy boundary intervals in objectives and constraints. *J. Energy Eng.* **2013**, *139*, 169–189. [CrossRef]
23. Xie, Y.L.; Li, Y.P.; Huang, G.H.; Li, Y.F. An interval fixed-mix stochastic programming method for greenhouse gas mitigation in energy systems under uncertainty. *Energy* **2010**, *35*, 4627–4644. [CrossRef]
24. Wu, C.B.; Huang, G.H.; Li, W.; Xie, Y.L.; Xu, Y. Multistage stochastic inexact chance-constraint programming for an integrated biomass-municipal solid waste power supply management under uncertainty. *Renew. Sustain. Energy Rev.* **2015**, *41*, 1244–1254. [CrossRef]
25. Golari, M.; Fan, N.; Jin, T.D. Multistage stochastic optimization for production-inventory planning with intermittent renewable energy. *Prod. Oper. Manag.* **2016**, *26*, 409–425. [CrossRef]
26. Fu, Z.H.; Wang, H.; Lu, W.T.; Guo, H.C.; Li, W. An inexact multistage fuzzy-stochastic programming for regional electric power system management constrained by environmental quality. *Environ. Sci. Pollut.* **2017**, *24*, 28006–28016. [CrossRef] [PubMed]
27. Wang, L.; Huang, G.H.; Wang, X.Q.; Zhu, H. Risk-based electric power system planning for climate change mitigation through multi-stage joint-probabilistic left-hand-side chance-constrained fractional programming: A Canadian case study. *Renew. Sustain. Energy Rev.* **2018**, *82*, 1056–1067. [CrossRef]

28. Fan, Y.; Huang, G.H.; Huang, K.; Baetz, B.W. Planning water resources allocation under multiple uncertainties through a generalized fuzzy two-stage stochastic programming method. *IEEE. Trans. Fuzzy Syst.* **2015**, *23*, 1488–1504. [CrossRef]

29. Aseeri, A.; Bagajewicz, M.J. New measures and procedures to manage financial risk with applications to the planning of gas commercialization in Asia. *Comput. Chem. Eng.* **2004**, *28*, 2791–2821. [CrossRef]

30. Ruszczyński, A.; Shapiro, A. Optimization of risk measures. *Risk Insur.* **2004**, *10*, 119–157.

31. Chen, C.; Li, Y.P.; Huang, G.H. An inexact robust optimization method for supporting carbon dioxide emissions management in regional electric-power systems. *Energy Econ.* **2013**, *40*, 441–456. [CrossRef]

32. Xie, Y.L.; Huang, G.H.; Li, W.; Ji, L. Carbon and air pollutants constrained energy planning for clean power generation with a robust optimization model—A case study of Jining City, China. *Appl. Energy* **2014**, *136*, 150–167. [CrossRef]

33. Huang, G.H.; Baetz, B.W.; Patry, G.G. A grey linear programming approach for municipal solid waste management planning under uncertainty. *Civ. Eng. Syst.* **1992**, *9*, 319–335. [CrossRef]

34. Huang, G.H. IPWM: An interval parameter water quality management model. *Eng. Optim.* **1996**, *26*, 79–103. [CrossRef]

35. Yu, C.S.; Li, H.L. A robust optimization model for stochastic logistic problems. *Int. J. Prod. Econ.* **2000**, *64*, 385–397. [CrossRef]

36. Leung, S.C.H.; Tsang, S.O.S.; Ng, W.L.; Wu, Y. A robust optimization model for multi-site production planning problem in an uncertain environment. *Eur. J. Oper. Res.* **2007**, *181*, 224–238. [CrossRef]

37. Huang, G.H.; Loucks, D.P. An inexact two-stage stochastic programming model for water resources management under uncertainty. *Civ. Eng. Environ. Syst.* **2000**, *17*, 95–118. [CrossRef]

38. Zibo City Bureau of Statistics. *Zibo City Statistical Yearbook 2014*; China Statistical Press: Beijing, China, 2015.

energies

MDPI

Article

Optimal Energy Management of Building Microgrid Networks in Islanded Mode Considering Adjustable Power and Component Outages

Van-Hai Bui [1], Akhtar Hussain [1], Hak-Man Kim [1,*] and Yong-Hoon Im [2,*]

[1] Department of Electrical Engineering, Incheon National University, 12-1 Songdo-dong, Yeonsu-gu, Incheon 406840, Korea; buivanhaibk@inu.ac.kr (V.-H.B.); hussainakhtar@inu.ac.kr (A.H.)

[2] Korea Institute of Energy Research, 152 Gajeong-ro, Yuseong-gu, Daejeon 34129, Korea

* Correspondence: hmkim@inu.ac.kr (H.-M.K.); iyh@kier.re.kr (Y.-H.I.); Tel.: +82-32-835-8769 (H.-M.K.); +82-42-860-3327 (Y.-H.I.); Fax: +82-32-835-0773 (H.-M.K.); +82-42-860-3098 (Y.-H.I.)

Received: 17 August 2018; Accepted: 5 September 2018; Published: 6 September 2018

Abstract: In this paper, an optimal energy management scheme for islanded building microgrid networks is proposed. The proposed building microgrid network comprises of several inter-connected building microgrids (BMGs) and an external energy supplier. Each BMG has a local combined heat and power (CHP) unit, energy storage, renewables and loads (electric and thermal). The external energy system comprises of an external CHP unit, chillers, electric heat pumps and heat pile line, for thermal energy storage. The BMGs can trade energy with other BMGs of the network and can also trade energy with the external energy supplier. In order to efficiently utilize the components of the BMGs and the network, the concept of adjustable power is adopted in this study. Adjustable power can reduce the operation cost of the network by increasing/decreasing the power of dispatchable units. In addition, the failure/recovery of components in the BMGs and the external system are also considered to analyze the performance of the proposed operation method. In order to optimally utilize the available resources during events, precedence among loads of BMGs and the external energy supplier is considered. Simulation results have proved the applicability of the proposed method for both normal islanded mode and with outage/recovery of equipment during the operation horizon. Finally, sensitivity analysis is carried out to analyze the impact of change in components' parameters values on the saved cost of the network.

Keywords: building microgrid; component outage; energy network; islanded microgrids; microgrid operation; re-optimization and rescheduling

---

1. Introduction

Microgrids are considered as a practical solution for increasing the service reliability, reducing the emissions and enhancing the energy utilization efficiency. These objectives can be achieved due to the ability of the microgrids to island during events and their capability to sustain the penetration of renewables. In addition, due to the deployment of microgrids in the proximity of consumers, the waste heat generated by distributed generators during generation of electricity can also be utilized [1,2], that is, energy utilization efficiency can be increased. Small-scale combined heat and power (CHP) units are deployed to enhance the dispatchability and efficiency of total energy in the microgrids [3]. In traditional power plants, typically 30% of the fuel's available energy is converted into usable energy, that is, electricity. Combined cooling, heat and power (CCHP)-based generation units can enhance the utilization of available energy to 75–80% [4]. Therefore, CHP/CCHP-based systems are of particular value for microgrids.

Among various other types of microgrids, building microgrids (BMGs) have gained popularity in the recent years due to large consumption of energy, especially heating and cooling energy [5–7].

The energy consumption of commercial and residential buildings is about 40% in the USA and 48% of this energy is used for heating and cooling purposes [5]. Similarly, the heating and cooling energy is about 47% of the net energy in operational buildings in China [6]. In EU also, buildings consume 40% of the energy and 79% of this energy is used for heating and cooling, mainly heating [7]. Therefore, the concept of building microgrids with local generations is emerging. In order to enhance the self-sufficiency of buildings, 10% of renewables-based energy is made an obligation for public buildings in Korea [8]. The self-sufficiency of buildings can be further increased by interconnecting several buildings, especially buildings having different consumption patterns. Therefore, various studies are conducted to evaluate the feasibility of interconnecting heterogeneous occupancy buildings [9–11].

In Reference [9], maximum utilization of solar energy is considered by interconnecting heterogeneous occupancy buildings while improving the thermal comfort of the residents. Different demand type prosumer buildings are considered in Reference [10] and seasonal demand variations are considered to optimize the energy management of the network. Similarly, a cooperative network of residential buildings is considered in Reference [11] and fluctuations in renewables and flexibilities in temperature requirement are exploited. The hourly fluctuations of heat and power demands in building microgrids are considered and optimal resources are determined in Reference [12]. A multi-objective operation model for building microgrids is proposed in Reference [13] by considering both operation cost and the comfort of the occupants. A central controller is proposed in Reference [14] for global optimal energy management of cooperative building microgrids with a comprehensive communication network. In addition, various studies on modeling and sizing of CHP/CCHP equipment for cooperative building microgrids [15,16], testbeds for evaluating the performance of energy networks [17] and potentials of CHP/CCHP systems in specific locations (UK [18], China [19], USA [20] and EU [21]) are also available.

Most of the studies available in the literature on a network of interconnected building microgrids [15–17] have focused on the grid-connected mode operation. However, the islanded mode operation is more challenging due to limited resources and inability to exchange power with the utility grid. Therefore, both grid-connected and islanded mode operation of building microgrids are considered in Reference [22–24]. However, a central energy management system (EMS) is utilized in these studies and failure of equipment is not considered. The central EMS needs an extensive communication infrastructure and may cause single point failure problems, in case of events. Meanwhile, various hierarchical EMSs are suggested in the literature [25–27] to overcome the problems of centralized EMSs. Similarly, the islanded mode operation becomes more challenging, if some of the equipment is also out-of-service due to any event. The islanded mode operation becomes more challenging in this case due to the coupling of thermal and electrical energies in the BMGs. The equipment failure is not considered in most of the existing studies in the literature.

In order to overcome the drawbacks of the existing literature, the islanded mode operation of a building microgrid network is considered in this study. This paper is an extension of the authors' previous work [10], where the grid-connected operation mode was analyzed. Similar to [10], a network of BMGs with an external energy supplier (EES) is considered. The BMGs have local energy sources and they can exchange energy among themselves as well as they can trade energy with the EES. A hierarchical two-level EMS is adopted to carry out the optimization of the proposed BMG network. The building EMSs (BEMSs) are responsible for energy management of the individual BMGs while the community EMS (C-EMS) is responsible for the operation of the entire network. In order to enhance the service reliability and reduce the operation cost, the concept of adjustable power is utilized in this study. Where, adjustable power is the amount of power increased by a BMG having cheaper generation sources to send the excess power to other BMG(s) having expansive generation sources and vice versa. In addition, the impact of equipment failure in the BMGs and the EES on the operation of the BMG network is analyzed. Priorities are defined for local and community loads to ensure the continuity of service to more critical loads during event cases when the available resources are not sufficient to fulfill all the load demands.

2. System Configuration and Operation Strategy of the Proposed CCHP System

2.1. CCHP System Configuration

In this paper, the configuration of the CCHP system is based on a pilot system in Korea Institute of Energy Research (KIER), Korea. The configuration of the proposed CCHP system is divided into two main parts, that is, energy networks and communication network, as shown in Figure 1a,b, respectively. The energy networks include electric, heating and cooling networks. In the electric network, the load demand in each building is initially fulfilled by local CHPs, renewable distributed generators (RDGs) and battery energy storage system (BESS). All buildings are interconnected and are also connected with an external system, which comprises of an external CHP (ECHP), electric heat pump (EHP), adsorption chiller (AC), heat pipeline (HPL) system and pump system. The whole system is operated in islanded mode. Therefore, all buildings cannot trade power with the utility grid. However, the BMGs can trade power with other BMGs of the network and can also trade with the EES. Similarly, the BMGs can trade heat energy with other BMGs of the network or with the HPL system. The HPL system plays an important role in heat network for charging/discharging heat energy to/from all the buildings. It also charges heat energy from the ECHP unit and discharges heat energy to AC for generating cooling energy. The BMGs fulfill their cooling load demand by buying cooling energy from the external energy network (EEN). The EEN generates cooling energy by utilizing EHP and/or AC units. In order to operate the CCHP system, the energy balance is maintained during all the operation intervals for electric, heating and cooling energies.

The communication network is shown in detail in Figure 1b. The communication network is designed as a hierarchical EMS, which includes three BEMSs and a C-EMS. The BEMSs are responsible for the optimal operation of all local resources in their respective buildings. The C-EMS gathers the information from all BEMSs and all components in the EES for optimal operation of the whole system. The detailed operation strategy of the CCHP system is presented in the next section for both normal and emergency operation modes. Throughout the paper, normal operation refers to islanded mode operation and emergency operation refers to operation with the outage of any equipment in the system.

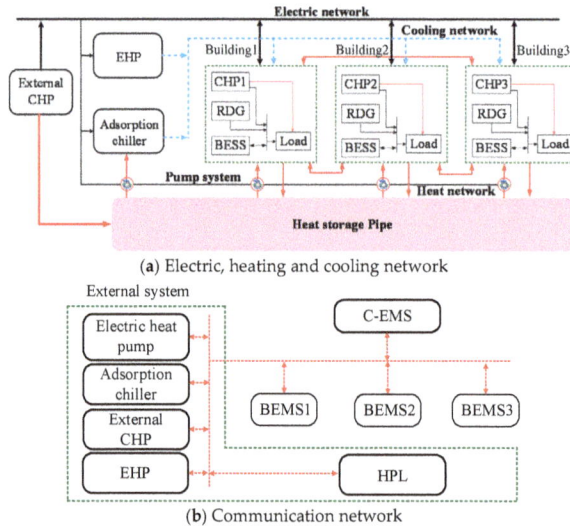

(a) Electric, heating and cooling network

(b) Communication network

Figure 1. CCHP system configuration.

2.2. Operation Strategy for CCHP System

Figure 2 presents the operation strategy of the proposed CCHP system in normal operation mode. Firstly, each BEMS gathers the information of all components in the building as an input and performs optimization to minimize the operation cost of the building. After performing the optimization, the amount of surplus/shortage energy and the output power of CHP units are determined. Based on the output power of CHP units, the bounds of the adjustable power of these CHP units are calculated. The information of surplus/shortage energy and the adjustable power are informed to the C-EMS. After receiving all the information from BEMSs and the components in the external system, the C-EMS performs central optimization. In this step, the amount of adjustable power, amount of energy trading among buildings and the amount of energy trading with the EES are determined. The optimal results are informed to the external system and all the BEMSs. Each BEMS reschedules the operation of all its component in the building. The output power of the CHP units is updated based on the amount of adjusted power. The detailed amount of power trading among buildings and the external system are also determined. Finally, the total operation cost of the building is determined and the BEMS sends the final operation schedules to all the components for implementation. The flowchart in Figure 2 shows the step-by-step optimization for day-ahead scheduling of the CCHP system. However, several events might occur in the CCHP system during the operation period, which could affect the operation of the CCHP system.

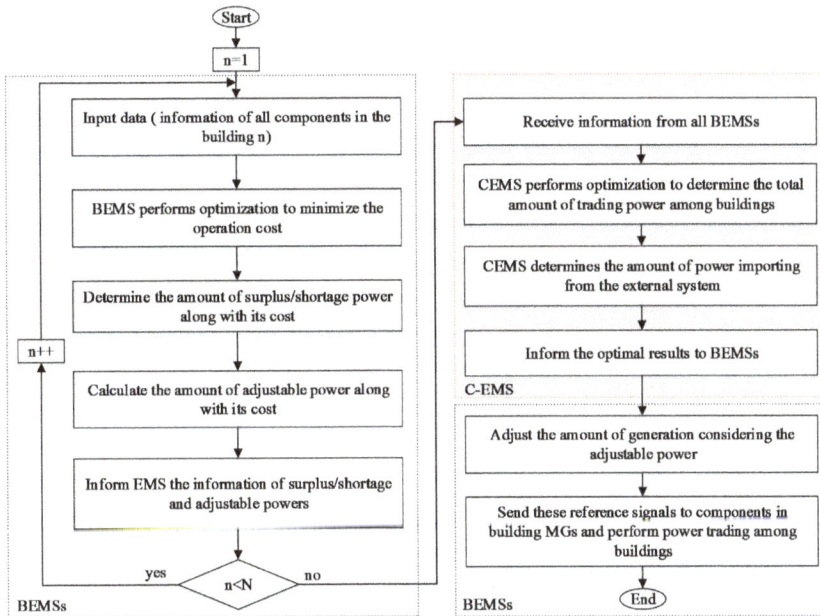

Figure 2. Operation strategy of the CCHP system in normal operation mode.

The operation of the CCHP system in an emergency case is presented in Figure 3. By using the proposed strategy, whenever an event occurs, the event location and event time are detected by the energy management system. The operation of the CCHP system is rescheduled based on the event location. If the event is inside the building, the BEMS reschedules the operation of all components from the event time to the end of the day and informs the C-EMS with the updated information. Similarly, the C-EMS reschedules the operation of the whole system based on the updated information. On the other hands, if the event is in the external system, the C-EMS detects the event and reschedules the system and informs the updated information to all BEMSs for rescheduling.

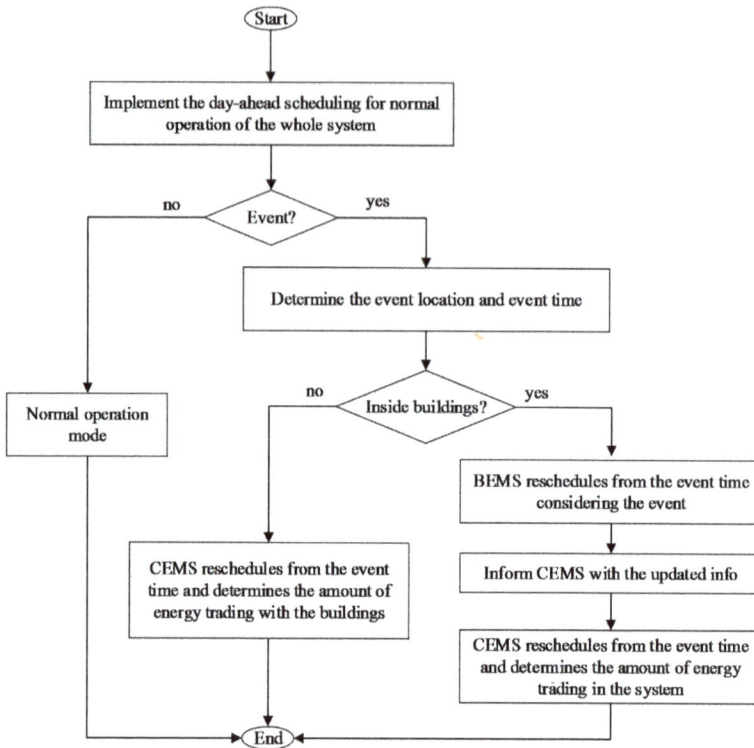

Figure 3. Operation strategy of the CCHP system in event time.

Figure 4 shows the rescheduling horizon for the CCHP system in case of events. The scheduling horizon for normal mode is 24 h (T) and operation is based on the day-ahead model. However, in case of an event at time h, the BEMSs and the C-EMS will switch their operation mode to emergency mode. The scheduling horizon of the emergency mode is from the event time $t = h$ to the end of the scheduling horizon (T). Besides, if the system has any storage system, the SOC also need to be updated for the rescheduling operation. In this way, the operation of the system can be updated based on the event occurrence time.

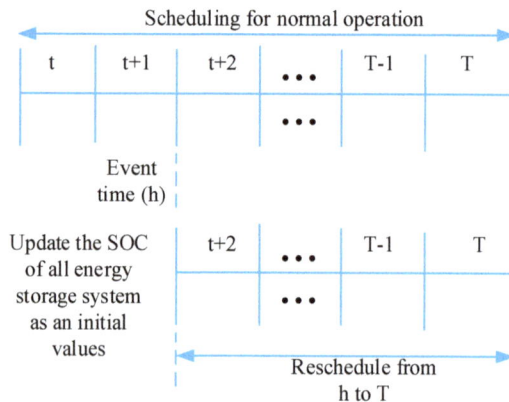

Figure 4. Rescheduling of the CCHP system with an event time.

3. Mathematical Model of CCHP System

In this section, a mixed integer linear program (MILP)-based three-step optimization model is presented for minimizing the operation cost of the entire CCHP system. In the first step, each BEMS optimizes the operation of all components inside the building. The surplus/shortage and the bounds of adjustable amount are proposed to the C-EMS. The C-EMS gathers all information from BEMSs as well as from the components of the EES and performs optimization for the whole system. Then the total amount of power sharing among buildings and with the external system is determined in this step. The results from C-EMS are informed to all buildings for rescheduling. In the third step, each BEMS reschedules its local resources considering the amount of adjustable power, power-sharing with other buildings and updates its operation cost. The detailed formulation is presented in the following sections.

3.1. Step 1: Local Optimization by BEMSs

In this step, each BEMS optimizes the operation of all components in each building for the islanded mode. The objective function is to minimize the operation cost of the building, as shown in Equation (1). The first term of Equation (1) represents the operation, start-up and shut-down costs of each CHP in the building. The second term represents the penalty based on the amount of shortage of power in the building. The third term shows the cost of trading heat energy with the external system. The last term shows the cost of buying cooling energy from the external system.

$$
\begin{aligned}
Min \sum_{t=1}^{T} \sum_{i=1}^{I} & \left(C_i^{CHP} \cdot P_{i,t}^{CHP} + x_{i,t} \cdot C_i^{SU} + y_{i,t} \cdot C_i^{SD} \right) \\
& + \sum_{t=1}^{T} \left(C_t^{Pen} \cdot P_t^{Short} \right) + \sum_{t=1}^{T} \left(PR_t^{BuyH} \cdot H_t^{Short} - PR_t^{SellH} \cdot H_t^{Sur} \right) + \sum_{t=1}^{T} \left(PR_t^{BuyC} \cdot Co_t^{Short} \right)
\end{aligned}
\tag{1}
$$

The objective Equation (1) is constrained by Equations (2)–(16). The upper and lower bounds of each CHP unit are given by Equation (2). Based on the on-off status of each CHP, the start-up and shut-down statuses are determined, as shown in Equations (3) and (4), respectively. The heat energy output of CHP is determined using the power-to-heat ratio, as given by Equation (5).

$$
u_{i,t} \cdot P_{i,Min}^{CHP} \leq P_{i,t}^{CHP} \leq P_{i,Max}^{CHP} \cdot u_{i,t} \quad \forall i \in I, t \in T \tag{2}
$$

$$
x_{i,t} = \max\{(u_{i,t} - u_{i,t-1}), 0\}, u_{i,t} \in \{0,1\} \quad \forall i \in I, t \in T \tag{3}
$$

$$
y_{i,t} = \max\{(u_{i,t-1} - u_{i,t}), 0\}, u_{i,t} \in \{0,1\} \quad \forall i \in I, t \in T \tag{4}
$$

$$
H_{i,t}^{CHP} = P_{i,t}^{CHP} \cdot rat_i^{CHP} \quad \forall i \in I, t \in T \tag{5}
$$

In each building, the energy balance between the supply and the demand is maintained during each operation interval. Equation (6) shows the power balance in a building. The power supply from RDG, CHPs and BESS should be equal to the power demand. The surplus/shortage power is traded either with the external system or with other buildings. Similarly, the heating and cooling energy balance are given by Equations (7) and (8), respectively. Due to the absence of cooling sources in individual buildings, the cooling demand is fulfilled by importing the cooling energy from the EES.

$$
P_t^{RDG} + \sum_{i=1}^{I} P_{i,t}^{CHP} + P_t^{Short} - P_t^{Sur} + P_t^{BDis} - P_t^{BChar} = P_t^{Load} \quad \forall t \in T \tag{6}
$$

$$
\sum_{i=1}^{I} H_{i,t}^{CHP} + H_t^{Short} - H_t^{Sur} = H_t^{Load} \quad \forall t \in T \tag{7}
$$

$$
Co_t^{Short} = Co_t^{Load} \quad \forall t \in T \tag{8}
$$

In the islanded mode, in order to reduce the load shedding amount, BESSs are used to shift the surplus power from off-peak load intervals to peak load intervals. The constraints for operation of BESS are shown in Equations (9)–(12). The bounds of charging and discharging power are determined in each interval based on the state of charge (SOC) of BESS at the end of the previous interval, as given by Equations (9) and (10), respectively. After performing charging/discharging, SOC of BESS is updated based on the amount of power charged/discharged and is carried to the next interval, as shown in Equation (11). The SOC of BESS is maintained within its capacity limits, as shown in Equation (12).

$$0 \leq P_t^{BChar} \leq P^{BCap} \cdot \left(1 - SOC_{t-1}^B\right) \cdot \frac{1}{1 - L^{BChar}} \quad \forall t \in T \tag{9}$$

$$0 \leq P_t^{BDis} \leq P^{BCap} \cdot SOC_{t-1}^B \cdot (1 - L^{BDis}) \quad \forall t \in T \tag{10}$$

$$SOC_t^B = SOC_{t-1}^B - \frac{1}{P^{BCap}} \cdot \left(\frac{1}{1 - L^{BDis}} \cdot P_t^{BDis} - P_t^{BChar} \cdot (1 - L^{BChar})\right) \quad \forall t \in T \tag{11}$$

$$0 \leq SOC_t^B \leq 1 \quad \forall t \in T \tag{12}$$

After performing the optimization, the bounds of increaseable and decreaseable electric/heat energy (i.e., adjustable bounds) are determined for each CHP unit based on its generated power by Equations (13) and (14), respectively. Finally, the amount of surplus, shortage and adjustable bounds are informed to the C-EMS for the second step of optimization.

$$P_{i,t}^{Avail\_inc} = P_{i,Max}^{CHP} - P_{i,t}^{CHP} \quad \forall i \in I, t \in T \tag{13}$$

$$H_{i,t}^{Avail\_inc} = H_{i,Max}^{CHP} - H_{i,t}^{CHP} \quad \forall i \in I, t \in T \tag{14}$$

$$P_{i,t}^{Avail\_dec} = P_{i,t}^{CHP} - P_{i,Min}^{CHP} \quad \forall i \in I, t \in T \tag{15}$$

$$H_{i,t}^{Avail\_dec} = H_{i,t}^{CHP} - H_{i,Min}^{CHP} \quad \forall i \in I, t \in T \tag{16}$$

3.2. Step2: Optimization for the Whole System with C-EMS

In step 2, the objective function aims to minimize the operation cost of the whole system, as shown in Equation (17). The first line of Equation (17) represents the operation cost of the ECHP unit considering start-up/shut-down costs and the cost of adjusting the output power of CHP units in the buildings. The second line of Equation (17) represents the total cost for importing power from the buildings and the profit gained by selling electric power from ECHP to the buildings. The third line of Equation (17) represents the cost for trading heat energy among buildings and with the heat pipeline system. The last term of Equation (17) represents the profit gained by selling the cooling energy to buildings to fulfill their cooling demands.

$$
\begin{aligned}
Min \sum_{t=1}^{T} &\left(C^{ECHP} \cdot P_t^{ECHP} + z_t \cdot C_{ECHP}^{SU} + k_t \cdot C_{ECHP}^{SD}\right) + \sum_{t=1}^{T}\sum_{n=1}^{N}\left(C_t^{Adj} \cdot \left(P_{n,t}^{Inc} - P_{n,t}^{Dec}\right)\right) \\
&+ \sum_{t=1}^{T}\left(PR_t^{Buy} \cdot \left(P_{EHP,t}^{Buy} + P_{AC,t}^{Buy} + P_{Pum,t}^{Buy}\right)\right) - \sum_{t=1}^{T}\left(PR_t^{SellE} \cdot P_t^{Esell}\right) \\
&+ \sum_{t=1}^{T}\left(PR_t^{BuyH} \cdot \sum_{n=1}^{N} H_{n,t}^{HPLChar} - PR_t^{SellH} \cdot \left(\sum_{n=1}^{N} H_{n,t}^{HPLDis} + H_{AC,t}^{HPLDis}\right)\right) \\
&- \sum_{t=1}^{T}\left(PR_t^{SellC} \cdot \sum_{n=1}^{N} Co_{n,t}^{Sell}\right)
\end{aligned}
\tag{17}
$$

The objective function in this step is also constrained by several constraints, as given by Equations (18)–(34). The operation bounds of ECHP, the start-up status and shut-down status are determined based on the on-off mode of ECHP, as shown in Equations (18)–(20), respectively. The heat output of the ECHP is determined by Equation (21) based on its power output and power-to-heat ratio.

The amount of heat generated by the ECHP is firstly charged to the HPL system and is only wasted if the HPL is fully charged, as shown in Equation (22).

$$v_t \cdot P_{Min}^{ECHP} \leq P_t^{ECHP} \leq P_{Max}^{ECHP} \cdot v_t \quad \forall t \in T \tag{18}$$

$$z_t = \max\{(v_t - v_{t-1}), 0\}, v_t \in \{0,1\} \quad \forall t \in T \tag{19}$$

$$k_t = \max\{(v_{t-1} - v_t), 0\}, v_t \in \{0,1\} \quad \forall t \in T \tag{20}$$

$$H_t^{ECHP} = P_t^{ECHP} \cdot rat^{ECHP} \quad \forall t \in T \tag{21}$$

$$H_t^{ECHP} = H_{ECHP,t}^{HPLChar} + H_{ECHP,t}^{Waste} \quad \forall t \in T \tag{22}$$

The power generated by ECHP can be used for EHP, AC and pump system or can be sold to the buildings, as given by Equation (23). The constraints related to the cooling energy balance are given by Equations (24) and (25). The EHP and AC are utilized for generating cooling energy and the cooling energy is sold to the buildings for fulfilling their cooling demands. The generation bounds of EHP and AC are presented by Equations (26) and (27), respectively. The total electric power utilized for EHP, AC and pump system is given by Equations (28)–(30). The power can either be received from ECHP or can be bought from other buildings.

$$P_t^{ECHP} = P_t^{Esell} + P_{ECHP,t}^{EHP} + P_{ECHP,t}^{AC} + P_{ECHP,t}^{Pum} \quad \forall t \in T \tag{23}$$

$$\left(P_{ECHP,t}^{EHP} + P_{EHP,t}^{Buy}\right) \cdot n_{E-C} + H_{AC,t}^{HPLDis} \cdot n_{H-C} = \sum_{n=1}^{N} Co_{n,t}^{Sell} \quad \forall t \in T \tag{24}$$

$$Co_{n,t}^{Sell} = Co_{n,t}^{Short} \quad \forall n \in N, t \in T \tag{25}$$

$$\left(P_{ECHP,t}^{EHP} + P_{EHP,t}^{Buy}\right) \cdot n_{E-C} \leq Co_{Max}^{EHP} \quad \forall t \in T \tag{26}$$

$$H_{AC,t}^{HPLDis} \cdot n_{H-C} \leq Co_{Max}^{AC} \quad \forall t \in T \tag{27}$$

$$P_t^{EHP} = P_{ECHP,t}^{EHP} + P_{EHP,t}^{Buy} \quad \forall t \in T \tag{28}$$

$$P_t^{AC} = P_{ECHP,t}^{AC} + P_{AC,t}^{Buy} \quad \forall t \in T \tag{29}$$

$$P_t^{Pum} = P_{ECHP,t}^{Pum} + P_{Pum,t}^{Buy} \quad \forall t \in T \tag{30}$$

The electric and heat energy balance between the internal system (i.e., buildings) and the external system (EES) is presented in Equations (31) and (32) considering the amount of adjustable power. The amount of adjusted heat energy is calculated based on the amount of adjusted power, as shown in Equations (33) and (34).

$$P_t^{Esell} + \sum_{n=1}^{N} P_{n,t}^{Sur} + \sum_{n=1}^{N} P_{n,t}^{Inc} = \sum_{n=1}^{N} P_{n,t}^{Short} + \sum_{n=1}^{N} P_{n,t}^{Dec} + P_t^{Ebuy} \quad \forall t \in T \tag{31}$$

$$\sum_{n=1}^{N} H_{n,t}^{Sur} + \sum_{n=1}^{N} H_{n,t}^{Inc} + \sum_{n=1}^{N} H_{n,t}^{Dis} = \sum_{n=1}^{N} H_{n,t}^{Short} + \sum_{n=1}^{N} H_{n,t}^{Dec} + \sum_{n=1}^{N} H_{n,t}^{Char} \quad \forall t \in T \tag{32}$$

$$H_{n,t}^{Inc} = \sum_{i=1}^{I} (rat_{i,t}^{CHP} \cdot P_{i,t}^{Inc}) \quad \forall n \in N, t \in T \tag{33}$$

$$H_{n,t}^{Dec} = \sum_{i=1}^{I} (rat_{i,t}^{CHP} \cdot P_{n,t}^{Dec}) \quad \forall n \in N, t \in T \tag{34}$$

The HPL system is used for storing the excess of heat energy. The surplus heat energy can either come from the buildings or from the ECHP in the EES. The heat energy in the HPL system is utilized for operating AC to generate cooling energy or to sell to other buildings having shortage of heat energy. The constraints related to the operation of the HPL system are shown in Equations (35)–(38). The bounds for charging and discharging heat energy to/from the HPL are determined based on the SOC of HPL system at the previous interval, as shown in Equations (35) and (36), respectively. The SOC of the HPL system is updated by Equation (37) based on the charging/discharging amount and losses. The operation bounds of HPL system are given by Equation (38).

$$0 \le \sum_{n=1}^{N} H_{n,t}^{HPLChar} + H_{ECHP,t}^{HPLChar} \le H^{HPLCap} \cdot \left(1 - SOC_{t-1}^{HPL}\right) \cdot \frac{1}{1 - L^{HPLChar}} \quad \forall t \in T \qquad (35)$$

$$0 \le \sum_{n=1}^{N} H_{n,t}^{HPLDis} + H_{AC,t}^{HPLDis} \le H^{HPLCap} \cdot SOC_{t-1}^{HPL} \cdot \left(1 - L^{HPLDis}\right) \quad \forall t \in T \qquad (36)$$

$$SOC_t^{HPL} = SOC_{t-1}^{HPL} - \frac{1}{P^{HPLCap}} \cdot \left(\begin{array}{c} \frac{1}{1 - L^{HPLDis}} \cdot \left(\sum_{n=1}^{N} H_{n,t}^{HPLDis} + H_{AC,t}^{HPLDis}\right) \\ -\left(\sum_{n=1}^{N} H_{n,t}^{HPLChar} + H_{ECHP,t}^{HPLChar}\right) \cdot \left(1 - L^{HPLChar}\right) \end{array} \right) \quad \forall t \in T \quad (37)$$

$$0 \le SOC_t^{HPL} \le 1 \quad \forall t \in T \qquad (38)$$

3.3. Step 3: Rescheduling by BEMSs

In this step, BEMS reschedules the operation of all components in each building considering the information from the C-EMS. BEMSs update the output power of CHP units as well as the amount of energy trading with other BMGs and the EES. During peak load intervals, the load shedding could be performed for maintaining the power balance in the system. Finally, the operation cost of each building is also updated based on the real output of CHP units and the amount of energy traded with other buildings or the EES, as shown in Equation (39).

$$\begin{aligned}
Min \sum_{t=1}^{T} \sum_{i=1}^{I} & \left(C_i^{CHP} \cdot \left(P_{i,t}^{CHP} + P_{i,t}^{Inc} - P_{i,t}^{Dec}\right) + x_{i,t} \cdot C_i^{SU} + y_{i,t} \cdot C_i^{SD} \right) \\
& + \sum_{t=1}^{T} \left(PR_t^{BuyE} \cdot P_t^{Rec} - PR_t^{SellE} \cdot P_t^{Send} \right) + \sum_{t=1}^{T} \left(PR_t^{RecH} \cdot H_t^{Rec} - PR_t^{SendH} \cdot H_t^{Send} \right) \\
& + \sum_{t=1}^{T} \left(PR_t^{BuyH} \cdot H_t^{HPLDis} - PR_t^{SellH} \cdot H_t^{HPLChar} \right) \\
& + \sum_{t=1}^{T} \left(PR_t^{BuyC} \cdot Co_t^{Short} \right) + \sum_{t=1}^{T} \left(pen_t^{Shed} \cdot P_t^{Shed} \right)
\end{aligned} \qquad (39)$$

Similarly, the operation bounds of CHP units are presented by Equation (40). The energy balance for electric and heat energies are shown in Equations (41) and (42), respectively, considering the amount of adjusted power and the energy traded within the CCHP system. Additionally, the objective Equation (39) is also constrained by Equations (3)–(5) and (8)–(12).

$$u_{i,t} \cdot P_{i,Min}^{CHP} \le P_{i,t}^{CHP} + P_{i,t}^{Inc} - P_{i,t}^{Dec} \le P_{i,Max}^{CHP} \cdot u_{i,t} \quad \forall i \in I, t \in T \qquad (40)$$

$$P_t^{RDG} + \sum_{i=1}^{I} \left(P_{i,t}^{CHP} + P_{i,t}^{Inc} - P_{i,t}^{Dec} \right) + P_t^{Rec} - P_t^{Send} + P_t^{BDis} - P_t^{BChar} = P_t^{Load} - P_t^{Shed} \quad \forall t \in T \quad (41)$$

$$\sum_{i=1}^{I} \left(H_{i,t}^{CHP} + H_{i,t}^{Inc} - H_{i,t}^{Dec} \right) + H_t^{Rec} - H_t^{Send} + H_t^{HPLDis} - H_t^{HPLChar} = H_t^{Load} + \sum_{i=1}^{I} \left(H_{i,t}^{Waste} \right) \quad \forall t \in T \quad (42)$$

4. Numerical Results

4.1. Input Data

The test CCHP system is comprised of three buildings and an external system and it is operated in islanded mode, as shown in Figure 1. The analysis is conducted for a 24-h scheduling horizon with 1 h time intervals [8,9]. All numerical simulations are coded in Visual Studio 2010 and are solved using the MILP solver CPLEX (12.6) [28]. The load (electric, thermal and cooling) profiles of all the three BEMSs during a weekday are shown in Figure 5a–c, respectively. Figure 5d shows the output of RDGs in all the BMGs. Similarly, Figure 6a–d show the electric, thermal, cooling profiles and the output of RDGs in all the BMGs for a weekend day, respectively. The information of CDG units in BMGs and EES are tabulated in Table 1. Power-to-heat ratio of CHPs is varied under different load-ability to better utilize the heat and power. However, frequent change in power-to-heat ration can adversely affect the lifetime of the CHP equipment. Therefore, the power-to-heat ratio of CHPs is varied for different seasons of the years, not on daily basis [29]. Due to the above-mentioned problems, the power-to-heat ratio is kept same in this study. Finally, the information of BESSs and HPL is tabulated in Table 2.

Table 1. Parameter of CHPs and ECHP.

Parameters	CHP1	CHP2	CH3	ECHP
Min. (kWh)	0	0	0	0
Max. (kWh)	1000	1000	1000	850
Operation cost (won/kWh)	95	100	80	70
Start-up cost (won)	200	200	200	300
Shut-down cost (won)	200	200	200	300
Ratio	3.5	1	1.8	3

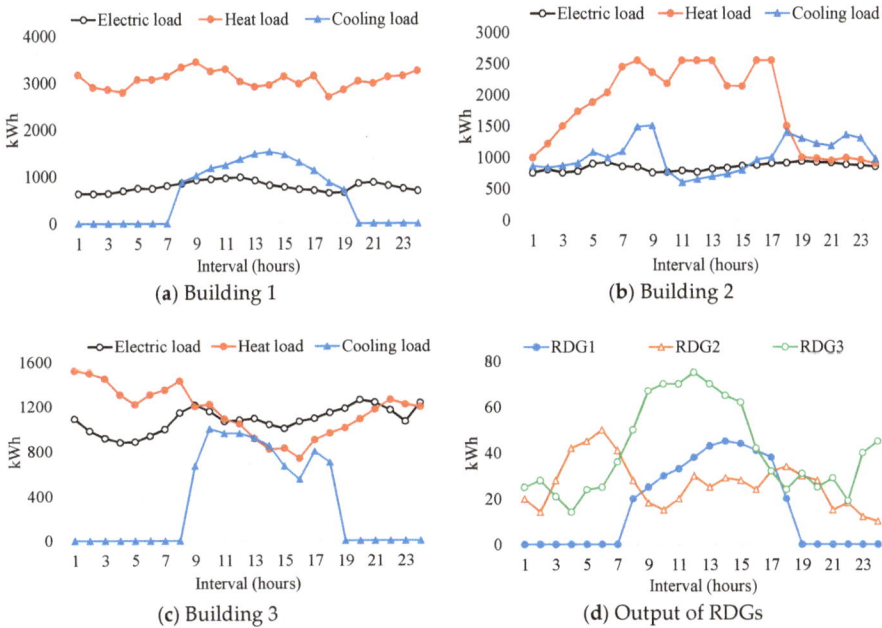

(a) Building 1

(b) Building 2

(c) Building 3

(d) Output of RDGs

Figure 5. Load profiles and output power of RDG in each building during a weekday.

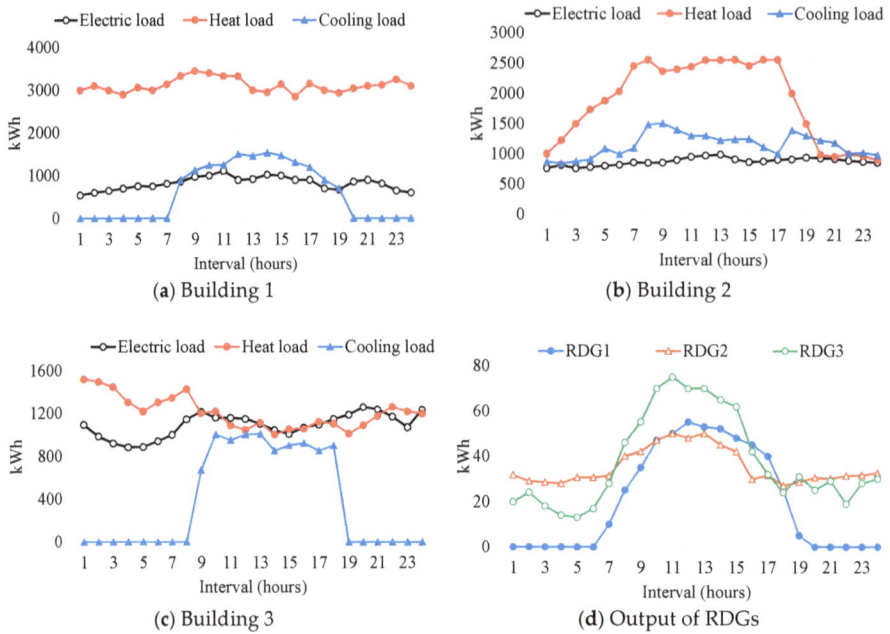

Figure 6. Load profiles and output power of RDG in each building during a weekend day.

Table 2. Parameter of BESSs and HPL system.

Parameters	BESS1	BESS2	BESS3	HPL
Min. (kWh)	0	0	0	2000
Max. (kWh)	200	250	300	50,000
Initial (kWh)	50	100	100	10,000
C-Loss (%)	5	5	5	5
D-Loss (%)	5	5	5	5

4.2. Operation of the CCHP System in Normal Operation

This section presents the normal operation of the CCHP system in islanded mode. The numerical results show the operation of the whole system during a weekday. In the first step, BEMSs optimize the operation of all component inside the buildings and inform the optimal results to the C-EMS. The operation of local resources depends on the operation cost of each CHP unit, power-to-heat ratio and the load profiles. In building 1, due to the lower operation cost of CHP, it always generates maximum power. The amount of shortage electric and heat energies are zeros during all the operation intervals. The surplus of electric/heat energy is traded with other buildings and/or the external system, as shown in Figure 7a. The operation cost of CHP in building 2 is higher along with a lower power-to-heat ratio. Therefore, there is no surplus of electric or heat energies. The CHP unit is only set to fulfill the local electric load, the amount of heat shortage is fulfilled by receiving heat from the HPL system and other buildings, as shown in Figure 7b. Similarly, in building 3, the CHP unit is also set to fulfill the local electric load as shown in Figure 7c. However, during peak load intervals (8, 17–14), the local sources cannot fulfill all the load. The information about shortage of power and surplus of heat is sent to the C-EMS. Due to the absence of local cooling sources, all the cooling load information is sent to the C-EMS as the amount of cooling shortage. The information on the surplus/shortage energies in each building is summarized in Figure 7.

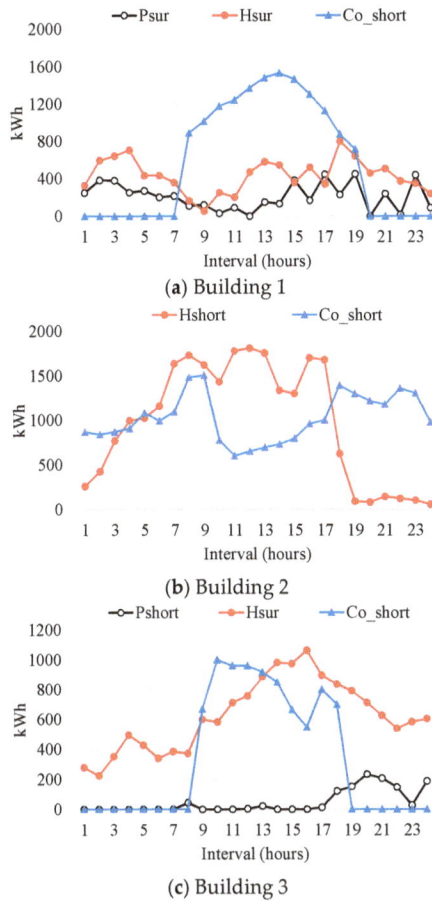

Figure 7. Information of the surplus/shortage energies in each building during a weekday.

After performing optimization by BEMSs, the output power of CHP units is determined in all the buildings. The bounds of adjustable power for electric and heat energies can be calculated by using Equations (13)–(16). Finally, BEMSs combine all information of surplus/shortage and adjustable energies and inform to the C-EMS.

The C-EMS gathers all information from BEMSs and performs optimization for minimizing the operation cost of the whole system. The operation of all components in the external system is summarized in Figure 8. The operation cost of ECHP is low, therefore, it is always set to generate maximum output power. The power is sent to all equipment in the external system, that is, AC, pumps and EHP. The surplus power is sold to the buildings, as shown in Figure 8a. However, if EHP requires a large amount of power, the external system also imports power from the buildings (interval 17), as shown in Figure 8b. Figure 8c shows that the output of AC and EHP is utilized to fulfill the cooling load demand of the BMGs. Figure 8d shows the amount of heat trading in the whole system. The main heat energy source for HPL is the ECHP. Whenever the ECHP is operated, the heat energy is charged to the HPL system. The HPL also receives heat energy from buildings (intervals 17–20). The HPL discharges heat energy to AC for generating cooling energy, which is used for fulfilling the cooling load of the buildings. Similarly, HPL can be discharged to feed the heating load demand of buildings having shortage of heat energy, that is, intervals 5–13, 16 and 17.

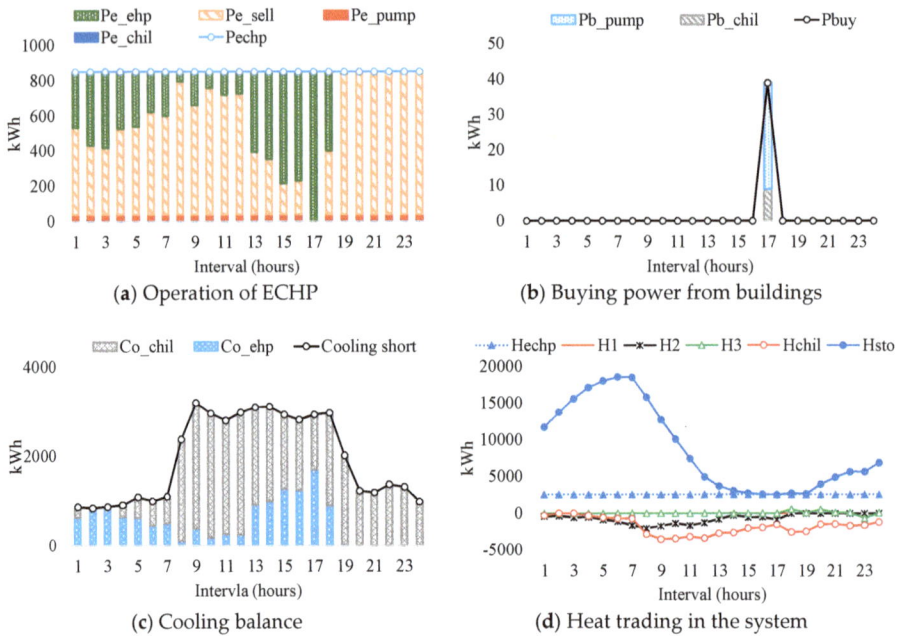

(a) Operation of ECHP

(b) Buying power from buildings

(c) Cooling balance

(d) Heat trading in the system

Figure 8. Operation of all components in the external system during a weekday.

Figure 9 shows the amount of adjusted power in each building, the amount of power traded with the external system and the amount of heat energy traded among BMGs. It can be observed from Figure 9a that the CHP units having high operation cost decrease their output power for reducing the operation cost. The shortage of electric and heat energies due to reduction in the amount of power of the CHP unit is imported from other buildings and/or the external system. For instance, at interval 1, CHP unit in building 2 decreases 740 kWh as shown in Figure 9a. The amount of decreased power is fulfilled by importing the cheaper power from building 1 (250 kWh) and the external system (490 kWh), as shown at interval 1 in Figure 9b. The amount of traded heat energy is depicted in Figure 9c. The surplus of heat energies from building 1 and 3 are sent to building 2 to reduce the amount of heat trading with the HPL system.

Finally, in order to analyze the impact of the adjustable power on the operation cost of the whole system, a comparison between two cases is considered for both a weekday and a weekend day. The first case considers the proposed method considering the adjustable power while the second case is without the adjustable power. It can be observed from Table 3 that by using the proposed method with the adjustable power, the operation cost of the system has reduced by 6.38% and 4.79% for a weekday and a weekend day, respectively, that is, cost savings. This reduction was due to decrease in the generation of CHP units having higher operation cost while increasing the output of CHPs in other BMGs having lower generation cost to fulfill the energy demand of the network. In a weekend day, the load demand is usually increased for residential buildings (building 2) and shopping malls (building 3) as compared to a weekday. However, the load demand in hospital building (building 1) changes randomly depending on the number of patients present in that particular day. During weekend day, the local generator in each building generated more power to fulfil all the local load demand. Therefore, the amount of surplus or adjustable power is decreased, which results in decreased power sharing among the BMGs of the network. As a result, the saved cost on a weekend day is lower than a weekday, as shown in Table 3.

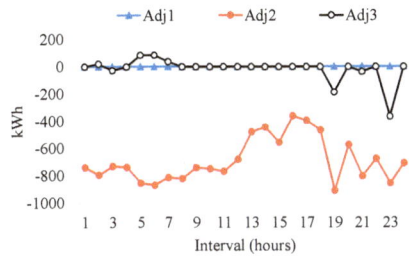

(a) Adjusted power

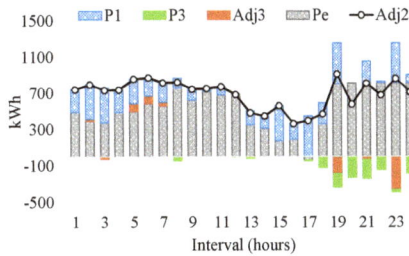

(b) Power trading with the external system

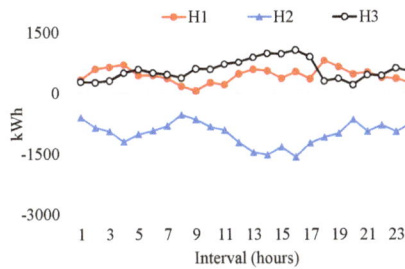

(c) Heat energy trading among buildings

Figure 9. Adjustable power and energy trading in buildings during a weekday.

Table 3. Saved cost with the proposed operation strategy.

Saved Cost	Weekday		Weekend Day	
	KRW	**%**	**KRW**	**%**
Method				
Without adjustable power	0	0	0	0
With adjustable power	417,655	6.38%	330,572	4.79%

4.3. Operation of the CCHP System during Events

The energy management systems, that is, BEMSs and C-EMS are designed to operate the CCHP system in both normal and emergency modes. Whenever any event occurs in the system, the system detects the event along with the occurrence time of the event and reschedules the operation of the whole system. In this study case, three event cases are analyzed in the CCHP system.

In the first event, the CHP unit in building 2 is out of service at interval 6. In the second event, the CHP unit in building 2 is recovered at interval 13. In the third event, the ECHP is out of service at interval 20. The detailed analysis of each event case is presented in the following section.

4.3.1. CHP Unit in Building 2 is Out of Service at Interval 6 and Recovered at Interval 13

In the first event, the BEMS of building 2 reschedules the operation schedule of all components in the building from event time (interval 6) to the end of the day (interval 24) considering fault in the CHP unit. The shortage of electric, heating and cooling energies are updated for building 2 as shown in Figure 10a. After receiving the updated information from BEMSs, C-EMS reschedules the operation of the whole system with the updated information. In this case, the surplus of electric and heat energies from other buildings are sent to building 2 for fulfilling the shortage power and reducing the amount of load shedding. For instance, at interval 6, the amount of shortage of power in building 2 is 850 kWh as shown in Figure 10c. However, the total surplus available from building 1 and imported power from the external system is 750 kWh. In this case, the building 3 can increase its generation amount by 100 kWh and send to building 2 for fulfilling the load as shown in Figure 10b. However, during peak load intervals (12–18, 20, 22, 24), the amount of shortage of power is very high. The load shedding is inevitable to maintain the power balance in the system, as shown in Figure 10c.

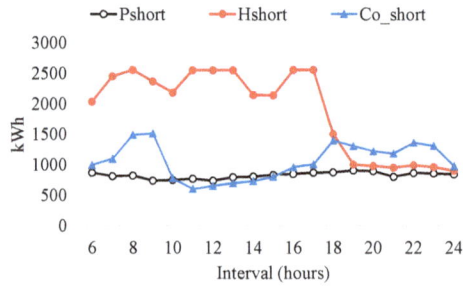

(a) Shortage/surplus energies from building 2

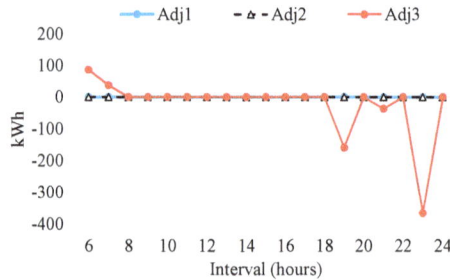

(b) Adjusted power

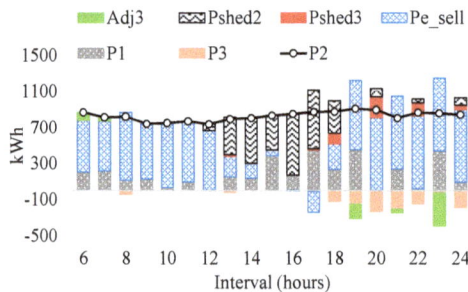

(c) Power trading with the external system

Figure 10. Rescheduling for the event case at interval 6 during a weekday.

At interval 13, the CHP unit is recovered from the fault. In the same way, BEMS of building 2 reschedules the operation of all the components considering the recovered component from interval 13 to interval 24. Figure 11a shows the surplus/shortage power information in building 2 after rescheduling. It can be observed that the shortage of electric energy is reduced to zero due to the recovery of CDG in BMG2. Similarly, the shortage of heat energy is also significantly reduced for the same reason. The updated information with the bounds of adjustable power is informed to the C-EMS for rescheduling the CCHP system. After rescheduling by C-EMS, it can be observed from Figure 11b that the amount of surplus power from the building 1 and the amount of importing power from the ESS are used for decreasing the output power of the expensive CGD unit (CDG2). The load shedding in building 2 and 3 are also recovered as shown in Figure 11c. From interval 13, the operation of the system is recovered and it operates similar to the normal operation mode, as shown in the previous section.

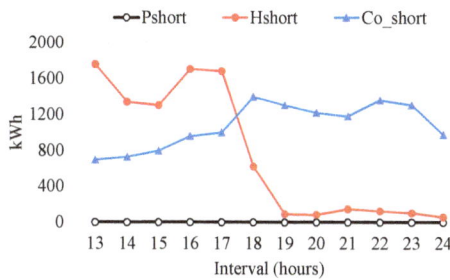

(a) Shortage/surplus energies from building 2

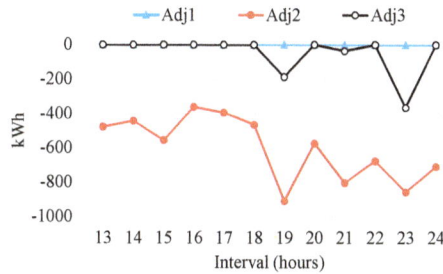

(b) Adjusted power

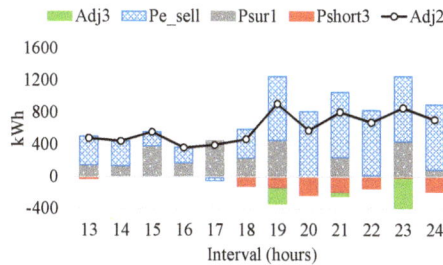

(c) Power trading with the external system

Figure 11. Rescheduling for the event case at interval 13 during a weekday.

4.3.2. ECHP Unit is Out of Service at Interval 20

In this case, the event is detected by C-EMS. Therefore, BEMSs are not required to reschedule at the first step. C-EMS reschedules the operation of the whole system. Due to the failure of ECHP, the external system imports power from the buildings to fulfill all the loads. The external loads are more important loads, that is, EHP, AC and pump system, due to their role in maintaining the service availability for the whole network. The interruption of service to such loads could have a significant effect on the operation of the whole system. For example, if the pump system is stopped, all the buildings and AC cannot trade heat energy from HPL system. Similarly, if the EHP or AC is interrupted, it could make a large amount of cooling load shedding in the whole system. Therefore, in this event, the buildings should provide the power to fulfill the all load demand of the EES. It can be observed from Figure 12a that the amount of buying power is enough for fulfilling all the load demands of the EES. In order to fulfill the loads in the external system, the local CHP units in the buildings increase the output power and send to the external system, as shown in Figure 12b. During intervals 20 and 22, the power supply is not enough for all the load in buildings. The power is sent to the high priority loads in the external system and the load shedding is performed in the buildings for maintaining power balance, as shown in Figure 12c.

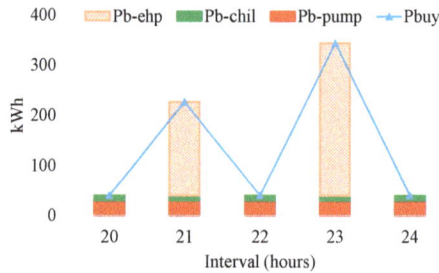

(a) Buying power from building

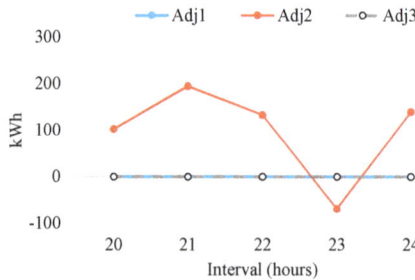

(b) Adjusted power

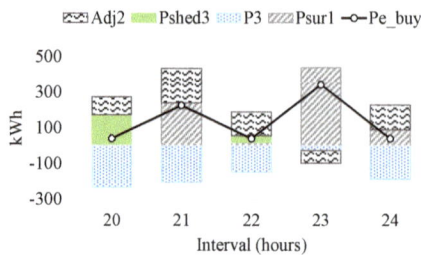

(c) Power trading with the external system

Figure 12. Rescheduling for the event case at interval 20 during a weekday.

4.3.3. Sensitivity Analysis

In this section, sensitivity analysis of different parameter values, that is, the operation cost of CHPs, the operation cost of ECHP, the initial value of BESSs and the initial value of HPL is presented to show the effect on the saving cost for a weekday. In order to analyze the effect of each parameter, the value of each parameter is varied by ±1%, ±2%, ±3%, ±4% and ±5%, individually while keeping the value of other parameters same.

In normal operation, the ECHP usually operates at maximum output power due to the cheaper operation cost, as shown in the previous sections. The surplus power from ECHP is sold to other buildings for reducing the output power of the expensive CHPs. Figure 13a,b shows the effect of the change in operation cost of CHPs in BMGs and ECHP on the saved cost, respectively. It can be observed that both the change in the operation cost of CHPs and ECHP have a significant effect on the saved cost. If the operation cost of CHPs is increased, the saved cost is increased because the buildings can receive power from ECHP and reduce more expensive CHPs in this case, as shown in Figure 13a. On the other hand, the increase in the operation cost of ECHP leads to the reduction of the saved cost because the surplus power from ECHP is sold to other building with a higher price, as shown in Figure 13b. The change in the initial value of BESSs and HPL have a minute effect on the saved cost. The change in initial values of BESSs have lesser impact on the saving cost a as compared to the change in operation cost of the CHPs. In addition, if the initial values of BESSs are increased, the output power of CHPs is slightly reduced during some intervals and the saved cost is also slightly increased, as shown in Figure 13c. Similarly, the initial value of HPL also have a lower effect on the saved cost. However, the capacity and the initial value of HPL is much higher than the BESSs. Thus, it has a more prominent effect on the saved cost as compared to the BESSs, as shown in Figure 13d.

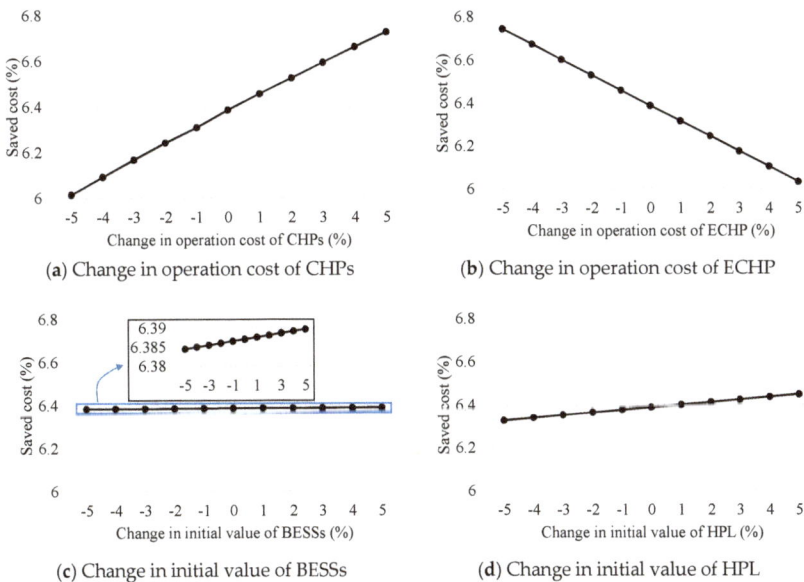

(a) Change in operation cost of CHPs

(b) Change in operation cost of ECHP

(c) Change in initial value of BESSs

(d) Change in initial value of HPL

Figure 13. Sensitivity analysis of different values of parameters on saved cost during a weekday.

5. Conclusions

An optimal energy management system for a building microgrid network in islanded mode is proposed in this study. The building microgrid network comprises of three buildings and an external energy supplier, which comprises of ECHP, EHP, chiller and HPL. The BMGs can trade energy among themselves as well as with the external energy network. In order to better utilize the resources of the network, the concept of adjustable power is adopted in this study. By using the proposed adjustable power, the operation cost has been reduced by 6.4% and 4.79%, respectively for a weekday and a weekend day in the tested network. The proposed adjustable power method can reduce the operation cost in normal operation and can reduce the load-shedding amount during outage events. In addition, the outage of equipment in the BMGs and the external energy supplier is also considered to analyze the performance of the proposed operation method. During outage events, power is increased by healthy BMGs and energy is shared with the on-emergency BMGs to minimize the load-shedding amount. Higher precedence is given to the loads responsible for continuing the service to the whole network, community-level resources in the external energy supplier over loads of individual BMGs. Simulation results have proved that the proposed method can optimally reschedule the available resources of the network upon outage/recovery of system components. The component values are changed by ±5% and a maximum of 6.74% saved cost is achieved for −5% reduction in the operation cost of the ECHP.

Author Contributions: V.-H.B. has designed the experiments and wrote the paper partially; A.H. conceived the idea and partially wrote the paper; H.-M.K. revised and analyzed the results; Y.-H.I. has revised the paper and helped in the idea finalization.

Funding: This work was supported by In-house Research and Development Program of the Korea Institute of Energy Research (KIER) (B8-2412).

Acknowledgments: This work was supported by In-house Research and Development Program of the Korea Institute of Energy Research (KIER) (B8-2412).

Conflicts of Interest: The authors declare no conflict of interest.

Nomenclature

Sets

N	Set of buildings
T	Set of time intervals in the scheduling horizon.
I	Set of CHPs

Indices

n	Index of buildings, running from 1 to N.
t	Index of time intervals, running from 1 to T.
i	Index of generators, running from 1 to I.

Parameters

C_i^{CHP}	Operation cost of CHP unit i
C_i^{SU}, C_i^{SD}	Start-up and shut-down costs of CHP unit i
C_t^{Pen}	Penalty for shortage power at t
rat_i^{CHP}	Power-to-heat ratio of CHP unit i
$PR_t^{BuyE}, PR_t^{SellE}$	Electricity buying/selling price signal for trading power at t
$PR_t^{RecH}, PR_t^{SendH}$	Heat buying/selling price signal for receiving/sending heat energy among BMGs at t
$PR_t^{BuyH}, PR_t^{SellH}$	Heat buying/selling price signal of BMGs for trading with the heat pipeline system at t
$PR_t^{BuyC}, PR_t^{SellC}$	Cooling buying/selling price signals for trading with the EES at t
P_t^{RDG}	Generation amount of renewable distributed generator at t
P_t^{Load}	Electric load amount at t
P^{BCap}	Capacity of BESS
H^{HPLCap}	Capacity of HPL
L^{BChar}, L^{BDis}	Charging/discharging losses of BESS
$L^{HPLChar}, L^{HPLDis}$	Charging/discharging losses of HP
H_t^{Load}	Heat load amount at t
Co_t^{Load}	Cooling load amount at t
C_t^{Adj}	Cost of adjustable power at t
C^{ECHP}	Operation cost of ECHP
C_{ECHP}^{SU}	Start-up cost of ECHP
C_{ECHP}^{SD}	Shut-down cost of ECHP
rat^{ECHP}	Power-to-heat ratio of ECHP

$Co_{Max}^{EHP}, Co_{Max}^{AC}$	Maximum cooling generation limit of EHP and AC
$P_{i,Min}^{CHP}, P_{i,Max}^{CHP}$	Minimum/Maximum power generation limits of CHP i at t
$H_{i,Min}^{CHP}, H_{i,Max}^{CHP}$	Minimum/Maximum heat generation limits of CHP i at t
$P_{Min}^{ECHP}, P_{Max}^{ECHP}$	Minimum/Maximum power generation limits of ECHP at t
n_{E-C}	Electricity to cooling ratio of EHP
n_{H-C}	Heat to cooling ratio of AC
pen_t^{Shed}	Penalty for load shedding at t

Variables

$P_{i,t}^{CHP}$	Amount of power generated by CHP unit i at t
$H_{i,t}^{CHP}$	Amount of heat generated by CHP i at t
$u_{i,t}, v_t$	Operation status of CHP i and ECHP at t
$x_{i,t}, y_{i,t}$	Start-up and shut-down status of CHP i at t
z_t, k_t	Start-up and shut-down status of ECHP at t
P_t^{ECHP}	Amount of power generated by ECHP at t
H_t^{ECHP}	Amount of heat generated by ECHP at t
$H_{ECHP,t}^{Waste}$	Amount of heat wasted by ECHP at t
P_t^{Short}, P_t^{Sur}	Amount of shortage/surplus power in a BMG at t
H_t^{Short}, H_t^{Sur}	Amount of shortage/surplus heat in a BMG at t
Co_t^{Short}	Amount of shortage of cooling in a BMG at t
P_t^{BChar}, P_t^{BDis}	Charging/discharging amount of BESS at t
SOC_t^B	State of charge of BESS at t
SOC_t^{HPL}	State of charge of HPL at t
$P_{i,t}^{Avail\_inc}, P_{i,t}^{Avail\_dec}$	Adjustable power for CHP unit i at t
$H_{i,t}^{Avail\_inc}, H_{i,t}^{Avail\_dec}$	Adjustable heat for CHP unit i at t
$H_t^{HPLChar}, H_t^{HPLDis}$	Heat charging/discharging amount to/from building at t
$H_{LCHP,t}^{HPLChar}, H_{AC,t}^{HPLDis}$	Amount of heat charged from large CHP and amount of heat discharged to AC at t
Co_t^{Buy}	Amount of cooling energy bought at t
$P_{n,t}^{Inc}, H_{n,t}^{Inc}$	Total increasable amount of electric/heat energies in building n at t
$P_{EHP,t}^{Buy}$	Amount of power bought from buildings for EHP at t
$P_{AC,t}^{Buy}$	Amount of power bought from buildings for AC at t
$P_{Pum,t}^{Buy}$	Amount of power bought from buildings for pumping system at t
$P_{ECHP,t}^{EPH}$	Amount of power received from ECHP for EHP at t
$P_{ECHP,t}^{AC}$	Amount of power received from ECHP for AC at t
$P_{ECHP,t}^{Pum}$	Amount of power received from ECHP for pumping system at t
P_t^{Esell}	Amount of power sold to the buildings at t from the external system
P_t^{Ebuy}	Amount of power bought from the buildings at t
$Co_{n,t}^{Sell}$	Amount of cooling energy sold to building n at t by the EES
P_t^{Rec}, P_t^{Send}	Amount of power received/sent by a BMG at t
H_t^{Rec}, H_t^{Send}	Amount of heat received/sent by a BMG at t
$H_{i,t}^{Waste}$	Amount of heat wasted by CHP unit i at t
P_t^{Shed}	Amount of load shed in a BMG at t

References

1. Mirez, J.; Hernandez-Callejo, L.; Horn, M.; Bonilla, L.M. Simulation of direct current microgrid and study of power and battery charge/discharge management. *DYNA* **2017**, *92*, 673–679.
2. Marnay, C.; Chatzivasileiadis, S.; Abbey, C.; Iravani, R.; Joos, G.; Lombardi, P.; Mancarella, P.; Appen, V.J. Microgrid evolution roadmap. In Proceedings of the 2015 International Symposium on Smart Electric Distribution Systems and Technologies (EDST), Vienna, Austria, 8–11 September 2015; pp. 139–144.
3. Gu, W.; Wu, Z.; Bo, R.; Liu, W.; Zhou, G.; Chen, W.; Wu, Z. Modeling, planning and optimal energy management of combined cooling, heating and power microgrid: A review. *Int. J. Electr. Pow. Energy Syst.* **2014**, *54*, 26–37. [CrossRef]
4. Cho, H.; Smith, A.D.; Mago, P. Combined cooling, heating and power: A review of performance improvement and optimization. *Appl. Energy* **2014**, *136*, 168–185. [CrossRef]

5. Carpenter, J.; Mago, P.J.; Luck, R.; Cho, H. Passive energy management through increased thermal capacitance. *Energy Build.* **2014**, *75*, 465–471. [CrossRef]

6. Li, H.; You, S.; Zhang, H.; Zheng, W.; Zheng, X.; Jia, J.; Ye, T.; Zou, L. Modelling of AQI related to building space heating energy demand based on big data analytics. *Appl. Energy* **2017**, *203*, 57–71. [CrossRef]

7. Energy Efficient Buildings: Europe. Available online: https://ovacen.com/wp-content/uploads/2014/09/edificios-energeticamente-eficientes-en-europa.pdf (accessed on 17 August 2018).

8. Oh, S.D.; Yoo, Y.; Song, J.; Song, S.J.; Jang, H.N.; Kim, K.; Kwak, H.Y. A cost-effective method for integration of new and renewable energy systems in public buildings in Korea. *Energy Build.* **2014**, *74*, 120–131. [CrossRef]

9. Korkas, C.D.; Baldi, S.; Michailidis, I.; Kosmatopoulos, E.B. Intelligent energy and thermal comfort management in grid-connected microgrids with heterogeneous occupancy schedule. *Appl. Energy* **2015**, *149*, 194–203. [CrossRef]

10. Hussain, A.; Bui, V.H.; Kim, H.M.; Im, Y.H.; Lee, J.Y. Optimal energy management of combined cooling, heat and power in different demand type buildings considering seasonal demand variations. *Energies* **2017**, *10*, 789. [CrossRef]

11. Ouammi, A. Optimal power scheduling for a cooperative network of smart residential buildings. *IEEE Trans. Sustain. Energy* **2016**, *7*, 1317–1326. [CrossRef]

12. Zidan, A.; Gabbar, H.A.; Eldessouky, A. Optimal planning of combined heat and power systems within microgrids. *Energy* **2015**, *93*, 235–244. [CrossRef]

13. Liu, G.; Ollis, T.B.; Xiao, B.; Zhang, X.; Tomsovic, K. Community Microgrid Scheduling Considering Network Operational Constraints and Building Thermal Dynamics. *Energies* **2017**, *10*, 1554. [CrossRef]

14. Dagdougui, H.; Ouammi, A.; Dessaint, L.; Sacile, R. Global energy management system for cooperative networked residential green buildings. *IET Renew. Power Gener.* **2016**, *10*, 1237–1244. [CrossRef]

15. Sun, T.; Lu, J.; Li, Z.; Lubkeman, D.; Lu, N. Modeling combined heat and power systems for microgrid applications. *IEEE Trans. Smart Grid* **2017**, *9*, 4172–4180. [CrossRef]

16. Zenginis, I.; Vardakas, J.; Abadal, J.; Echave, C.; Morato, M.; Verikoukis, C. Optimal power equipment sizing and management for cooperative buildings in microgrids. *IEEE Trans. Ind. Inform.* **2018**. [CrossRef]

17. Thangavelu, S.R.; Nutkani, I.U.; Hwee, C.M.; Myat, A.; Khambadkone, A. Integrated electrical and thermal grid facility-testing of future microgrid technologies. *Energies* **2015**, *8*, 10082–10105. [CrossRef]

18. Salem, R.; Bahadori-Jahromi, A.; Mylona, A.; Godfrey, P.; Cook, D. Comparison and Evaluation of the Potential Energy, Carbon Emissions and Financial Impacts from the Incorporation of CHP and CCHP Systems in Existing UK Hotel Buildings. *Energies* **2018**, *11*, 1219. [CrossRef]

19. Luo, Z.; Gu, W.; Sun, Y.; Yin, X.; Tang, Y.; Yuan, X. Performance Analysis of the Combined Operation of Interconnected-BCCHP Microgrids in China. *Sustainability* **2016**, *8*, 977. [CrossRef]

20. Anne, H.; Rick, T.; Michael, F.; Rachel, W. Combined Heat and Power (CHP) Technical Potential in the United States. Available online: https://www.energy.gov/sites/prod/files/2016/04/f30/CHP%20Technical%20Potential%20Study%203-31-2016%20Final.pdf (accessed on 17 August 2018).

21. Weber, M. Combined Heat and Power in Europe. University of Edinburgh 2012. Available online: http://www.csas.ed.ac.uk/__data/assets/pdf_file/0006/81285/Weber_on_CHP_V3_FINAL_120330.pdf (accessed on 17 August 2018).

22. Li, Z.; Xu, Y. Optimal coordinated energy dispatch of a multi-energy microgrid in grid-connected and islanded modes. *Appl. Energy* **2018**, *210*, 974–986. [CrossRef]

23. Kim, J.Y.; Park, J.H.; Lee, H.J. Coordinated control strategy for microgrid in grid-connected and islanded operation. *IFAC Proc. Vol.* **2011**, *44*, 14766–14771. [CrossRef]

24. Mehrasa, M.; Pouresmaeil, E.; Jørgensen, B.N.; Catalão, J.P. A control plan for the stable operation of microgrids during grid-connected and islanded modes. *Electr. Power Syst. Res.* **2015**, *129*, 10–22. [CrossRef]

25. Wang, Y.; Mao, S.; Nelms, R.M. On hierarchical power scheduling for the macrogrid and cooperative microgrids. *IEEE Trans. Ind. Informat.* **2015**, *11*, 1574–1584. [CrossRef]

26. Tian, P.; Xiao, X.; Wang, K.; Ding, R. A hierarchical energy management system based on hierarchical optimization for microgrid community economic operation. *IEEE Trans. Smart Grid* **2016**, *7*, 2230–2241. [CrossRef]

27. Bui, V.H.; Hussain, A.; Kim, H.M. A multiagent-based hierarchical energy management strategy for multi-microgrids considering adjustable power and demand response. *IEEE Trans. Smart Grid* **2018**, *9*, 1323–1333. [CrossRef]

28. *IBM ILOG CPLEX V12.6 User's Manual for CPLEX 2015, CPLEX Division*; ILOG: Incline Village, NV, USA, 2015. Available online: https://www.ibm.com/support/knowledgecenter/SSSA5P_12.6.2/ilog.odms.studio.help/pdf/usrcplex.pdf (accessed on 17 August 2018).

29. Liu, Y.; Gao, S.; Zhao, X.; Zhang, C.; Zhang, N. Coordinated operation and control of combined electricity and natural gas systems with thermal storage. *Energies* **2017**, *10*, 917. [CrossRef]

energies

MDPI

Article

Fuzzy Portfolio Optimization of Power Generation Assets

Barbara Glensk and Reinhard Madlener *

Institute for Future Energy Consumer Needs and Behavior (FCN), School of Business and Economics/E.ON Energy Research Center, RWTH Aachen University, Mathieustrasse 10, 52074 Aachen, Germany; BGlensk@eonerc.rwth-aachen.de
* Correspondence: RMadlener@eonerc.rwth-aachen.de; Tel.: +49-241-8049-820; Fax: +49-241-8049-829

Received: 14 October 2018; Accepted: 2 November 2018; Published: 6 November 2018

Abstract: Fuzzy theory is proposed as an alternative to the probabilistic approach for assessing portfolios of power plants, in order to capture the complex reality of decision-making processes. This paper presents different fuzzy portfolio selection models, where the rate of returns as well as the investor's aspiration levels of portfolio return and risk are regarded as fuzzy variables. Furthermore, portfolio risk is defined as a downside risk, which is why a semi-mean-absolute deviation portfolio selection model is introduced. Finally, as an illustration, the models presented are applied to a selection of power generation mixes. The efficient portfolio results show that the fuzzy portfolio selection models with different definitions of membership functions as well as the semi-mean-absolute deviation model perform better than the standard mean-variance approach. Moreover, introducing membership functions for the description of investors' aspiration levels for the expected return and risk shows how the knowledge of experts, and investors' subjective opinions, can be better integrated in the decision-making process than with probabilistic approaches.

Keywords: portfolio analysis; semi-mean-absolute deviation model; fuzzy set theory; optimal power generation mix

---

1. Introduction

The purpose of the portfolio selection problem is to find combinations of investment possibilities which best meet the objectives of the investor. This analysis needs various types of information and should be based on criteria which can provide some guidance about what is important and unimportant, or what is relevant and irrelevant. Although the weighting of these objectives and the criteria depend on the type of investor, the two that are common to all investors are expected return maximization and risk minimization. If investors are rational, they want the return to be high and prefer certainty to uncertainty. Moreover, the optimal portfolio enables the investor to mitigate risk and opportunities with respect to a wide range of alternatives.

The foundation of modern portfolio analysis was laid by Harry Markowitz in the middle of the 20th century [1]. He considered returns of assets as random variables and introduced the mean-variance approach. He identified the portfolio return as an expected return, which is a sum of the product between the asset's expected return and its shares in the portfolio, and the risk measured as the volatility (variance) of the (stochastic) value of the expected return. Furthermore, he assumed the multivariate normal distribution for the rates of return and the quadratic form for the investor's utility (preferences) function. The purpose of mean-variance portfolio (MVP) analysis is the maximization of the portfolio's expected return and the minimization of the portfolios's risk. Searching for efficient portfolios ("efficient" in this context means there exists no other portfolio with the same or a smaller variance that has a larger return, and no portfolio with the same or a larger return that has a smaller risk) could be conducted by solving one of two problems: (1) maximization of the portfolio's expected return by a given accepted risk level or (2) minimization of the portfolio's risk by some given required portfolio return level.

The methodology proposed by Markowitz has seen an extensive development since 1952 but also a lot of criticism. Trying to avoid some of the rigid assumptions of MVP analysis and to simplify the solution methodology, a number of alternative approaches have been proposed and applied. For example, the computational complexity connected with the mean-variance model (necessity of estimation of the variance and covariance matrix) led to the linearization of the objective function. Furthermore, the popularity among investors of other risk measures, such as mean absolute deviation (MAD), value at risk (VaR), expected shortfall (conditional value at risk—CVaR), or semi-variance, is growing.

The MAD model is one of the alternatives to the classic mean-variance model in which the measure of risk (variance) is replaced by the absolute deviation. Konno and Yamazaki [2] proposed the MAD model as a linear model for portfolio selection and tested its application on data from the Tokyo stock market. These authors observed that the MAD model can be used as an alternative to the Markowitz model because the calculated optimal portfolios and their performances are quite similar to each other. Moreover, a linear problem could be solved more easily than a quadratic one. Furthermore, the authors noticed that the MAD model could be used to tackle large-scale problems where a dense covariance matrix can occur, and that it does not require any specific type of return distribution. They also showed that the proposed method encompasses all properties of the MVP analysis. However, the method and its advantages were not widely appreciated in the financial engineering community and were criticized by statisticians. Konno and Koshizuka [3] reviewed some of the more important properties of the MAD portfolio optimization model. They pointed out that the MAD model is superior to the MVP model both theoretically and computationally, and that this model belongs to a class of mean-lower partial risk models, which are more adequate to problems with asymmetric return distributions. The MAD model proposed by Konno and Yamazaki [2] found interest among other researchers and applicaton to other financial markets. Application of the MAD model (based on the linear semi-mean-absolute deviation risk function) enabled Mansini and Speranza [4] to introduce new specifications derived from market structure and from operative constraints into the portfolio selection model of the Milan stock exchange. De Silva et al. [5] applied the MAD as well as the CVaR approach in order to avoid inefficient, low return, and/or high-risk portfolios on the Brazilian stock exchange. Furthermore, Liu [6] used the concept of the mean-absolute deviation function proposed by [2] when the asset returns from financial markets are represented by interval data. The author noticed that the ability to calculate the bounds of the investment return can help initiate wider applications in portfolio selection problems. A brief review of the variety of solvable linear programming portfolio optimization models presented in the literature, where several different risk measures (such as MAD or CVaR) were applied, can be found by Mansini et al. [7]. The authors discussed the relative and absolute form of these models and their applications.

The mean-variance portfolio selection model, and other existing portfolio selection models, are based on probability theory. However, as a number of empirical studies have shown, those probabilistic approaches only partly capture reality, in contrast to fuzzy sets theory. Fuzzy sets theory can be used for a better description of real systems (situations) that are very often uncertain and vague in different ways [8]. Zimmermann, in his seminal book [8], explains the vagueness, fuzziness, and uncertainty in real-world systems as well as the usefulness of the application of fuzzy sets theory in order to model uncertainty. He develops the formal framework of fuzzy mathematics and presents the survey of the most interesting applications of the theory.

Fuzzy sets theory and fuzzy logic, the latter of which is an extension of classic argumentation (conventional logic) to argumentation that is closer to humans, was introduced by Lotfi A. Zadeh [9]. In the classical set theory, objects can either belong to a set or not, and there are no intermediate steps of membership. In contrast, a fuzzy set also allows for blurred states. Zadeh ([9], p. 338) defines a fuzzy set as follows: *"A fuzzy set is a class of objects with a continuum of grades of membership. Such a set is characterized* by a membership (characteristic) function which assigns to each object a grade of membership ranging between zero and one". The extension of fuzzy logic is possibility theory, introduced also by Zadeh [10] and advanced by Dubois and Prade [11]. Zadeh tries to explore some of the elementary properties of the possibility distribution concept and explains the importance of possibility theory, which arises from the fact that the information for the decisions is possibilistic rather than probabilistic in nature. Moreover, possibility theory is not a substitute for probability theory but deals with another kind of uncertainty. In possibility theory, the fuzzy variables are associated with possibility distributions in a similar way as random variables are with probability distributions. In contrast to probability theory, the possibility distribution function is defined by a so-called 'membership function' which describes the degree of affiliation of fuzzy variables. Membership functions which is a fundamental part of fuzzy sets theory, allows the gradual assessment of the membership of elements into the set and can characterize the fuzziness, have different forms. The most popular ones used are triangular, trapezoidal, or parabolic. However, other membership functions are hyperbolic, inverse-hyperbolic, exponential, logistic, or piecewise-linear.

Application areas for fuzzy sets theory are broad, ranging from different control systems, engineering, and consumer electronics to business economics, including decision theory [12], or financial problems, such as portfolio selection [13]. By using a fuzzy approach, the knowledge of experts, investors' subjective opinions, but also quantitative and qualitative analysis, can be better integrated into decision problems. Wang and Zhu [14] and Fang et al. [13] give a survey of the progress made in recent years in the direction of fuzzy portfolio optimization. They present different portfolio selection models with fuzzy objectives and/or fuzzy constraints. One of the possibilities is that of using fuzzy numbers to define the coefficients of the objectives and constraints; another one is applying the so-called aspiration (or satisfaction) level. Another concept of fuzzy portfolio problems considers models with interval coefficients, where expected returns are treated as interval numbers, and where so-called pessimistic and optimistic satisfaction indices are introduced (for more information, see [15]). Furthermore, Tanaka and Guo [16] proposed the use of possibility distributions in order to model uncertainty in returns. They defined upper and lower possibility distributions which should reflect experts' knowledge with regard to the portfolio selection problem.

The aim of this paper is to present a portfolio selection model for energy utilities by employing alternative risk measures, such as semi-mean-absolute deviation, which is one of the first attempts at the linearization of portfolio selection models, and comparing it with the standard mean-variance approach. Moreover, the contribution of this paper is an application of fuzzy sets theory to portfolio optimization problems in combination with alternative portfolio risk measures that can be more adequate for portfolios of real assets (such as power plants). The argument is that, in the case of power generation assets, the distribution of the power plant's return measure and commodity prices as well as other parameters taken into consideration differ from the normal distribution assumed in the standard mean-variance portfolio approach, potentially causing biased results.

The remainder of this paper is organized as follows: Section 2 deals with the application of portfolio analysis to the energy sector and in particular to power generation mixes. In Section 3, fuzzy semi-mean-absolute deviation portfolio selection models and a "return" definition for power generation mixes are introduced. Section 4 shows an empirical example for the presented methodologies. Section 5 provides some conclusions.

2. Applications of Portfolio Analysis to Power Generation Assets

The energy utilities are confronted with a very diverse range of resource options in their energy planning, but also a dynamic, complex, and uncertain future. Financial investors are used to dealing with uncertainty and commonly evaluate such problems with portfolio theory. According to portfolio theory, they could choose a specific risk level and then aim to maximize the portfolio's return. Furthermore, the diversification of portfolio assets is the best means of hedging future risk. Therefore, mean-variance portfolio theorists have found a new application field in the energy sector, and the theory seems to be a well-suited complementary methodology to the problem of planning and evaluating power portfolios and strategies.

The first application of portfolio theory to the energy sector was presented by Bar-Lev and Katz [17]. They applied the Markowitz portfolio approach to optimize the fossil fuel mix for electric utilities in the US market and examined whether the power utilities are efficient users of fossil fuels. More specifically, they considered a two-dimensional optimization problem with fuel cost and risk minimization. More recent literature contains further applications of portfolio theory to energy markets in different countries or regions from a utility point of view (for a recent review, see [18]). For example, Awerbuch and Berger [19] used MVP analysis for the electricity market in the European Union. They proposed a more complex model where fuel costs and also operating and capital costs were explicitly considered. Their results indicate that the current EU electricity mix is sub-optimal from a risk-return perspective. Furthermore, they conclude that fixed-cost technologies, such as many of those based on renewables, must be a part of any efficient portfolio. Roques et al. [20] considered the portfolio problem based on the MVP methodology from a cost perspective, but also added revenues and included the net present value (NPV) of the investment. They analyzed energy markets in the UK, and concluded that the optimal portfolio consists of natural-gas-combined cycle (NGCC) and a few nuclear power plants. Further applications of portfolio theory to the energy sector can be found in [18,21]. In [21], the cost approach was again applied and electricity production costs considered, which included fuel costs together with operating, capital, and external costs. Moreover, for this study of the Swiss and the U.S. energy market, the authors adopted seemingly unrelated regression estimation (SURE). The Swiss power generation mix was also the goal of the analysis presented in [22]. They are the first to implement the NPV criterion for portfolio analysis, following the Markowitz model. Moreover, they explicitly differentiated between base-load and peak-load technologies. In the work of Borchert and Schemm [23], the application of the CVaR as a risk measure in portfolio analysis for wind power projects in Germany was presented. Glensk et al. [24] also applied CVaR as a risk measure, but instead of power generation assets they analyzed portfolios of contracts from the European Energy Exchange (EEX) and the Polish Power Exchange (POLPX). They further pointed out that the proposed approach can be useful, especially for retailers on both markets, but that the impact of negative energy prices, as found e.g., on the EEX, should also be investigated.

The studies mentioned above reflect a considerable and growing interest in applying portfolio analysis on (liberalized) energy markets. Moreover, they point out different definitions of return and risk used by authors applying MVP theory to power generation portfolios. Analogically to the financial markets, the decision-making process in energy planning is complex and multidimensional. Economic, social, and environmental aspects; technical parameters; and different risks have to be taken into account. Regarding all these different aspects, the risk connected to electricity price, fuel cost, carbon dioxide cost, operation and maintenance costs, capital cost, but also to the capacity factor of a power plant, affect the measure of return and thus also the decision-making process and outcome. Application of portfolio theory can help to eliminate these risks and to explain the complex interactions between these parameters.

When applying portfolio theory to power generation mixes, appropriate definitions of return and risk are needed. Project evaluation methods and measures (such as net present value, internal and modified internal rate of return, profitability index, payback or discounted payback time) commonly used in finance management could also be useful proxies for the construction of power generation mixes. Each of these measures give different pieces of relevant and valuable information needed in the decision-making process. However, the net present value and the internal rate of return are the most often ones used. According to the short literature review presented, the NPV criterion is already one useful profitability indicator for energy projects and power generation selection problems, next to the annual expected return.

3. Model Specification

In this section, we present the model formulation for the selection of the optimal power generation mix, considering the semi-mean-absolute deviation as a risk measure and assuming that each rate of return is a possibilistic variable.

3.1. Semi-Mean-Absolute Deviation (SMAD) Model

In general, the portfolio optimization problem is a two-dimensional optimization model, where portfolio risk is minimized and return is maximized. Most dissatisfaction with the variance introduced by Markowitz as a risk measure stems from the fact that it does not differentiate between gains and losses. Moreover, Harry Markowitz himself and William Sharpe, among other economists, acknowledge that the original modern portfolio theory formulation has important limitations ([25], p. 428): *"Under certain conditions, the mean-variance approach can be shown to lead to unsatisfactory predictions of behavior. Markowitz suggests that a model based on the semi-variance would be preferable; in light of the formidable computational problems; however, he bases his analysis on the variance and standard deviation."* As shown in [26], the mean-absolute deviation as a measure of variability is less sensitive to outliers, and equivalent to mean-variance under the assumption of the normal distribution of the returns. Furthermore, the author shows that the MAD model possesses several advantageous theoretical properties, such that all capital asset pricing model relations for the mean-variance model also hold for the MAD model. Moreover, the MAD model is more compatible with the fundamental principle of rational decision-making. Despite the fact that the mean-variance Markowitz (1952) model is a solid, well-known method, which laid the foundation for portfolio theory, it has been proven that MAD produces similar portfolio returns [5,27], which makes us confident in MAD's optimization ability. Considering this alternative definition of risk MAD, as proposed by Konno and Yamazaki [2], the portfolio rate of return, R_p, is given as:

$$R_p = E\left[\sum_{i=1}^{n} R_i x_i\right] = \sum_{i=1}^{n} E(R_i) x_i, \tag{1}$$

and defined as the expected value of a sum of the product between the assets' expected return, R_i, and its shares in the portfolio, x_i, and the portfolio risk, w_p, defined by the expected value of the mean absolute deviation between the realization of the portfolio's rate of return and its expected value, as:

$$w_p = E\left[\left|\sum_{i=1}^{n} R_i x_i - E\left[\sum_{i=1}^{n} R_i x_i\right]\right|\right]. \tag{2}$$

Regarding these definitions of portfolio return and risk, a two-dimensional optimization problem can be formulated as follows:

$$E\left[\left|\sum_{i=1}^{n} R_i x_i - E\left[\sum_{i=1}^{n} R_i x_i\right]\right|\right] \to \min,$$

$$\sum_{i=1}^{n} E(R_i) x_i \to \max, \tag{3}$$

$$\sum_{i=1}^{n} x_i = 1,$$

$$0 \le x_i \le x_{i,max},$$

where $x_{i,max}$ is the maximal share of asset i in the portfolio.

Assuming that the measure of risk is defined through the mean absolute deviation of the portfolio's rate of return below the average (for the investor, only the downside risk is problematic, and thus actually relevant) the semi-mean-absolute deviation portfolio selection model can be specified as:

$$E\left[\left|min\left\{0, \sum_{i=1}^{n} R_i x_i - E\left[\sum_{i=1}^{n} R_i x_i\right]\right\}\right|\right] \rightarrow min,$$

$$\sum_{i=1}^{n} E(R_i)x_i \rightarrow max, \qquad (4)$$

$$\sum_{i=1}^{n} x_i = 1,$$

$$0 \leq x_i \leq x_{i,max}.$$

The semi-mean-absolute deviation belongs to the favorite risk measure used in portfolio selection models. In contrast to the variance, the semivariance as well as the CVaR (mentioned in Section 1) fulfill all desirable properties of a "good" risk measure, and the SMAD satisfies two out of four properties of coherent risk measures (see [28]), such as positive homogenity and subadditivity [29,30]. Nevertheless, the standard deviation, mean absolute deviation as well as mean absolute lower and upper semi-deviation belong to the general deviation risk measures introduced by Rockafellar et al. [31,32] as an extension of standard deviation, but they need not to be symmetric with respect to the upside and downside value of the random variable. The authors developed a theory of deviation measures axiomatically, presenting key examples and tracing the relationships with concepts of coherent risk measures. The main possible advantage of mean absolute deviation and its downside version is the relation to linear programming computations of optimal portfolios. Moreover, as shown in [33,34], because the MAD is a symmetric measure and the absolute semi-deviation is its half.

Regarding the considerations presented by Konno and Koshizuka [3] and Speranza [35], the equivalent to the semi-mean-absolute deviation model in (4) is given as:

$$\frac{1}{T} \sum_{t=1}^{T} d_t \rightarrow min,$$

$$\sum_{i=1}^{n} E(R_i)x_i \rightarrow max,$$

$$d_t \geq -\sum_{i=1}^{n} (R_{it} - E(R_i)) x_i, \qquad (5)$$

$$d_t \geq 0,$$

$$\sum_{i=1}^{n} x_i = 1,$$

$$0 \leq x_i \leq x_{i,max}.$$

3.2. Fuzzy Semi-Mean-Absolute Deviation (FSMAD) Model

The imperfect knowledge about returns and the resulting uncertain environment imply that the use of precise mathematics to model a complex system is insufficient. In order to capture the complex reality of decision-making problems, fuzzy sets theory is proposed as an alternative to the commonly used probabilistic approach. Different studies about fuzzy portfolio selection show that different elements can be fuzzified. Some of them suggest the use of a possibility distribution to capture model uncertainty on returns, while others propose fuzzy formulations. In this paper, we present only some of them, with a specific focus on the application to the energy sector.

Assuming that each rate of return, R_i, is a possibilistic variable, the simple conversion of model (5) presented above is replacing the expected value of the rate of return for asset i, $E(R_i)$, by an adequate fuzzy mean value. This fuzzy mean value depends on the form of membership function assumed. Considering R_i as a triangular fuzzy variable determined by triplet (a, α, β) of crisp numbers with center a, left-width $\alpha > 0$ and right-width $\beta > 0$, the interval-valued possibilistic mean, $M_{tr}(R_i)$, proposed by Carlsson and Fullér [36], is given as:

$$M_{tr}(R_i) = a + \frac{\beta - \alpha}{6}. \tag{6}$$

Triangular membership functions are mostly used for fuzzy logic control problems. The motivation behind their utilization stems from its simplicity. However, for many other applications, such as decision-making problems, triangular membership functions are not appropriate, and using a trapezoidal membership function is more reasonable. For a trapezoidal membership function [36], the interval-valued possibilistic mean value, $M_{trap}(R_i)$, with a tolerance interval $[a, b]$ and left- and right-width α and β, respectively, is given by

$$M_{trap}(R_i) = \frac{a + b}{2} + \frac{\beta - \alpha}{6}. \tag{7}$$

Some applications of this approach on the financial markets are proposed and shown by [13,37–40], among others. These authors considered portfolio selection models with both the variance as a risk measure as well as the mean-absolute deviation.

Taking into consideration a trapezoidal membership function and its interval-valued possibilistic mean value, the SMAD portfolio selection model (5) has to be reformulated. Moreover, regarding one of the optimization approaches mentioned in Section 1, the risk minimization problem for a given desired portfolio return level, R_0, will be considered, and is specified as:

$$\frac{1}{T} \sum_{t=1}^{T} d_t \rightarrow \min,$$

$$\sum_{i=1}^{n} M_{trap}(R_i) x_i \geq R_0,$$

$$d_t + \sum_{i=1}^{n} \left(R_{it} - M_{trap}(R_i) \right) x_i \geq 0, \tag{8}$$

$$d_t \geq 0,$$

$$\sum_{i=1}^{n} x_i = 1,$$

$$0 \leq x_i \leq x_{i,max}.$$

Let us further propose that the return R_i is a fuzzy variable determined by a trapezoidal membership function with a tolerance interval $[a_i, b_i]$ and left- and right-width α_i and β_i, respectively $(R_i = (a_i, b_i, \alpha_i, \beta_i))$. Vercher et al. [38] assumed that the left and the right reference functions are all of the same shape and are presented as the linear combination, which expresses the total fuzzy return on a portfolio, as in:

$$R_p = \sum_{i=1}^{n} R_i x_i = \left(\sum_{i=1}^{n} a_i x_i, \sum_{i=1}^{n} b_i x_i, \sum_{i=1}^{n} \alpha_i x_i, \sum_{i=1}^{n} \beta_i x_i \right) =$$
$$= \left(R_{pa}(x), R_{pb}(x), R_{p\alpha}(x), R_{p\beta}(x) \right). \tag{9}$$

Regarding the semi-mean-absolute deviation portfolio selection model (4), the above-presented fuzzy portfolio return (9), and following deliberations presented by [38], the portfolio selection model is defined as an optimization problem, which, for a given required return level, R_0, yields a minimum-risk portfolio and which can be formulated as:

$$\sum_{i=1}^{n} \left(b_i - a_i + \frac{1}{3} (\alpha_i + \beta_i) \right) x_i \to \min,$$

$$\sum_{i=1}^{n} \left(\frac{1}{2} (a_i + b_i) + \frac{1}{6} (\beta_i - \alpha_i) \right) x_i \geq R_0, \tag{10}$$

$$\sum_{i=1}^{n} x_i = 1,$$

$$0 \leq x_i \leq x_{i,max}.$$

As mentioned before, investment decisions are generally influenced by uncertain or hardly predictable social and economic circumstances. In such cases, the application of optimization methods is not always the best in comparison to the satisfaction approach. Especially in portfolio selection problems, the investor always has certain objective values concerning the expected return and a certain degree of risk. In problems encountered in a real world, some vague aspiration level is based on experiences and knowledge of decision-makers. Therefore, it is more natural to denote an individual's aspiration level as a fuzzy number. Additionally, the concept of employing fuzzy numbers to express an investor's aspiration level is also connected with choosing an adequate membership function for model formulation. Watada [15] was first to propose the use of a trapezoidal membership function in order to describe the return and risk aspiration level. Based on the Bellman–Zadeh maximization principle [12], he introduced a two-dimensional Markowitz portfolio selection problem with a certain aspiration level as fuzzy numbers. However, he then also discovered that by employing this function, there exist some difficulties when solving the portfolio selection problem. In order to avoid these problems, Watada proposed a nonlinear logistic membership function, arguing that a logistic function is more appropriate to vague goal levels of investors and, in addition, that the trapezoidal function (considered at first) is an approximation of the logistic function.

Taking into account the proposition and assumptions introduced by Watada [15], the logistic membership function for the portfolio's expected return and risk are given by $\mu_R(R_p)$ and $\mu_w(w_p)$, respectively, as:

$$\mu_R(R_p) = \frac{1}{1 + \exp(-\alpha_R(R_p - R_M))}, \tag{11}$$

$$\mu_w(w_p) = \frac{1}{1 + \exp(\alpha_w(w_p - w_M))}, \tag{12}$$

where R_M and w_M are the mid-points (of the expected return and risk, respectively) at the membership function value $\lambda = 0.5$, and α_R and α_w determine the shape of the membership functions $\mu_R(R_p)$ and $\mu_w(w_p)$, respectively. The two parameters R_M and w_M are determined by $\frac{R_s + R_n}{2}$ and $\frac{w_s + w_n}{2}$ where R_s is sufficiency and R_n the necessity level for the return, and w_s is sufficiency and w_n the necessity level for the risk measure. The values of the sufficiency and necessity levels for return and risk can be provided by the decision-maker, although the well-known (and in this work employed) method is the one proposed by Zimmermann [41].

Considering the Bellman–Zadeh maximization principle [12], the general portfolio selection problem with aspiration levels can be formulated as follows:

$$\lambda \to \max,$$

$$\mu_R(R_p) \geq \lambda,$$

$$\mu_w(w_p) \geq \lambda,$$

$$\sum_{i=1}^{n} x_i = 1, \tag{13}$$

$$0 \leq x_i \leq x_{i,max},$$

$$\lambda \geq 0.$$

Reconsidering the semi-mean-absolute deviation portfolio selection model (5) and by replacing $\mu_R(R_p)$ and $\mu_w(w_p)$ in model (13) with Equations (11) and (12), respectively, the following problem results:

$$\lambda \to \max,$$

$$\lambda + \exp\left(-\alpha_R\left(\sum_{i=1}^{n} E(R_i)x_i - R_M\right)\right)\lambda \leq 1,$$

$$\lambda + \exp\left(\alpha_w\left(\frac{1}{T}\sum_{t=1}^{T} d_t - w_M\right)\right)\lambda \leq 1,$$

$$d_t + \sum_{i=1}^{n}(R_{it} - E(R_i))x_i \geq 0, \tag{14}$$

$$d_t \geq 0,$$

$$\sum_{i=1}^{n} x_i = 1,$$

$$0 \leq x_i \leq x_{i,max},$$

$$\lambda \geq 0.$$

The two exponential constraints in the above-presented model can be transformed into

$$\alpha_R\left(\sum_{i=1}^{n} E(R_i)x_i - R_M\right) \geq \log\frac{\lambda}{1-\lambda}, \tag{15}$$

$$-\alpha_w\left(\frac{1}{T}\sum_{t=1}^{T} d_t - w_M\right) \geq \log\frac{\lambda}{1-\lambda}. \tag{16}$$

Replacing $\Lambda = \log\frac{\lambda}{1-\lambda}$ in (15) and (16), the model formulation (14) is equivalent to:

$$\Lambda \to \max,$$

$$\alpha_R\sum_{i=1}^{n} E(R_i)x_i - \Lambda \geq \alpha_R R_M,$$

$$\alpha_w\frac{1}{T}\sum_{t=1}^{T} d_t + \Lambda \leq \alpha_w w_M,$$

$$d_t + \sum_{i=1}^{n}(R_{it} - E(R_i))x_i \geq 0, \tag{17}$$

$$d_t \geq 0,$$

$$\sum_{i=1}^{n} x_i = 1,$$

$$0 \leq x_i \leq x_{i,max},$$

$$\Lambda \geq 0.$$

According to the parameters α_R and α_w, which determine the shape of the membership functions, the aspiration levels can be described accurately. The calculation of the parameters α depends on the membership function used in the model. For the membership function presented in this work for the return (11) and risk (12), $\alpha_R = \frac{6.91}{0.5(R_s - R_n)}$ and $\alpha_w = \frac{6.91}{0.5(w_s - w_n)}$, respectively (for more information, see [42]). Therefore, the portfolio selection model presented is convenient for different investors and their individual investment strategies. The application of this model for onshore wind power plants can be found in [43].

3.3. Return Definition for the Power Generation Portfolio Selection Problem

As mentioned in Section 2 above, many of the differences in the applications of portfolio analysis to energy markets are connected with the choice of the selection criteria used (and particularly differing return and risk definitions). Power generation mix problems have their specific character connected with the evaluation of power plants. In an evaluation process, different measures and methods borrowed from finance can be used. The choice about which of them to adopt depends on the decision-maker and her/his expectations regarding portfolio analysis. By searching for an appropriate and useful proxy, not only costs and revenues should be taken into consideration, but also technical parameters and the expected (remaining) lifetimes of the power plants concerned. Furthermore, an important point in the analysis is the impact of new investments on the existing portfolio. The massive changes in worldwide energy industries are the results of a number of factors, such as the increase in energy demand, growing industrialization processes, environmental policy and resource limitations. Many countries and energy providers are obliged to reconstruct (renew, expand, etc.) their power generation mix and to develop new, more sustainable possibilities of producing energy (especially carbon-free or low-carbon technologies).

Regarding all these aspects, the NPV approach appears to be not a perfect one but a relatively suitable measure as a portfolio selection criterion (see, e.g., [20] or [22] where the NPV was applied for the selection of power generation assets) both for new investments and also for existing power plants. Another possibility is the use of the annual return as a selection criterion. In contrast to the NPV, the annual return is a static measure and requires that the forecasts of all factors are taken into consideration in the analysis (see, e.g., [44]). Following an approach similar to the one used in [20,22], the objective variable is defined as the NPV of the investments. A favorable characteristic of the NPV approach is the allowance of a risk- and time-adjusted discounting, thus respecting the time value of different investments. Especially in the power supply industry, where investment horizons often span several decades, this is a very important point. It is based upon the discounted cash-flow technique and given by:

$$NPV = \sum_{t=0}^{T} \frac{CF_t}{(1 + WACC)^t}, \tag{18}$$

where CF_t denotes the annual cash flow and the WACC (weighted average cost of capital) is used as the discount rate.

The estimation of cash flows for power plants during the operation time requires input data, such as revenues obtained from energy production, fuel costs, carbon dioxide mitigation costs, operation and maintenance costs, capital costs as well as the depreciation rate. The calculation of a cash flow starts with the determination of earnings before interest and taxes (EBIT), which is given as the difference between revenues and costs including depreciation. After tax subtraction, earnings before interest and after tax (EBIAT) are obtained. Finally, to achieve the cash flow, depreciation expenses are added to EBIAT (for more information, see e.g., [45]).

4. Results

This section presents several empirical applications of the above-described models for energy generation mixes. Considering power plants (in this case, existing power plants owned by E.ON in various location Germany and new investments announced by E.ON), we use the project's NPV (in €/kW) as a proxy for portfolio selection, and risk is defined as the semi-mean-absolute deviation from the expected return (also in €/kW). All necessary information about the power plants considered, as well as their economic and technical data needed for the NPV estimation undertaken by Monte Carlo Simulation (MCS) and using the ORACLE® Crystal Ball software (version 11.1), are reported in detail in [44,46], respectively.

The calculations of efficient portfolios were made for three different fuzzy set models: FSMAD Model 1 (= model (8)), FSMAD Model 2 (= model (10)), and FSMAD Model 3 (= model (17)). As an extension, a comparison of the presented results of these three models with the classic MV (mean-variance) model and with the SMAD (semi-mean-absolute deviation) model is also provided. The classic MV model was proposed by [1]; it is a portfolio selection model with the standard deviation used as a risk measure. This is a two-objective optimization problem, specified as:

$$\sum_{i=1}^{n} x_i^2 \sigma_i^2 + 2 \sum_{i=1}^{n-1} \sum_{j=i+1}^{n} x_i x_j \sigma_i \sigma_j \rho_{ij} \rightarrow \min,$$

$$\sum_{i=1}^{n} E(R_i) x_i \rightarrow \max,$$

$$\sum_{i=1}^{n} x_i = 1 \quad \text{and} \quad 0 \leq x_i \leq x_{i,max},$$

where σ_i^2 denotes the variance of component asset i, σ_i the volatility of asset i, and ρ_{ij} the correlation coefficient between i and j. Efficient frontiers were obtained through the implementation of linear and quadratic programming in the dynamic, object-oriented programming language Python 2.6.

4.1. FSMAD Model for Power Generation Portfolios

Figure 1a–c show the efficient frontiers for existing power plants and all announced new investments obtained by using the FSMAD models presented in Section 3. Figure 1a shows the efficient frontiers where the projects' return is considered as a possibilistic variable with the trapezoidal membership function and adequate fuzzy mean value. We observe that the increase of the return (NPV/installed capacity) is faster than the risk increase. Comparing Figure 1a with Figure 1b, where efficient frontiers are obtained by application of FSMAD Model 2, the shifting of the efficient frontiers can be noticed. The efficient portfolios depicted in Figure 1b for some specific return level have higher risk levels than the efficient portfolio presented in Figure 1a, which were obtained with another fuzzy approach to portfolio selection. Considering the investor's aspiration levels, FSMAD Model 3 was analyzed as well and the results are shown in Figure 1c. In this case, the set of efficient portfolios is significantly smaller than for the two models presented first. Furthermore, by comparing Figure 1a and Figure 1c, one can observe that the efficient portfolios obtained by using FSMAD Model 3 (Figure 1c) from the upper part of the efficient frontiers depicted in Figure 1a (FSMAD Model 1). From inspection of Figure 1d, this outcome becomes even more clear.

Tables A1–A6 in the Appendix show the exact compositions of the efficient portfolios presented in Figure 1a,c. The models analyzed in this work contain restrictions regarding the maximal share of each technology allowed in the power generation mix. Such restrictions are necessary from a technical point of view and ensure that all resulting portfolios are technically feasible.

Additional information obtained from this analysis is related to new investments taken under consideration. Especially in Figure 1a,b, in some cases, efficient frontiers for existing technologies and new investments are located above the efficient frontiers obtained for the existing technologies only. It means that including these investments in the existing power generation mix can apparently improve the efficiency of the portfolio.

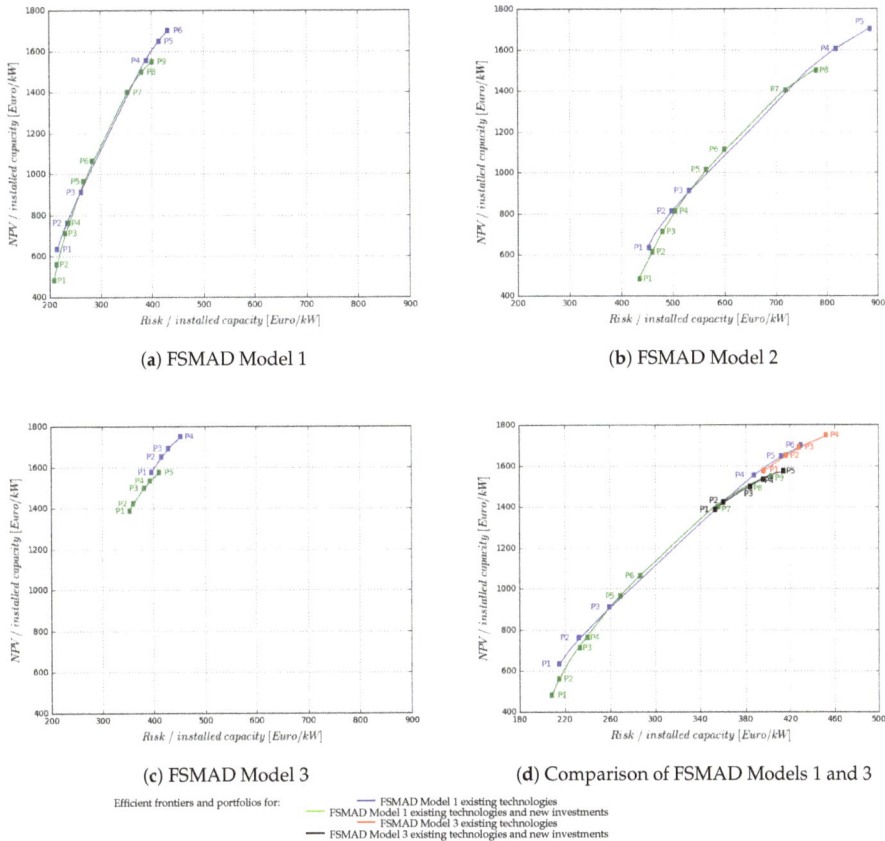

(a) FSMAD Model 1

(b) FSMAD Model 2

(c) FSMAD Model 3

(d) Comparison of FSMAD Models 1 and 3

Efficient frontiers and portfolios for:
— FSMAD Model 1 existing technologies
— FSMAD Model 1 existing technologies and new investments
— FSMAD Model 3 existing technologies
— FSMAD Model 3 existing technologies and new investments

Figure 1. Efficient frontiers obtained for different FSMAD models.

4.2. Comparative Analysis

The second part of our analysis shows the comparison of results obtained between the fuzzy portfolio selection models with the well-known mean-variance portfolio selection model proposed by Markowitz, and with the SMAD model, the latter of which is used as a benchmark model for our study as well.

Figure 2 presents the location of the efficient frontiers obtained for FSMAD Model 1 and the SMAD Model. The first conclusion is that both models give more or less the same results. The very similar shape and position of the curves can be explained by the very similar arithmetic and fuzzy mean values obtained by Monte Carlo simulation. Nevertheless, closer inspection shows that the efficient frontier printed in blue (FSMAD Model 1) is located slightly above the efficient frontier printed in red (SMAD Model) and in green (FSMAD Model 1 regarding new investments); it is also located slightly above the frontier printed in black (SMAD Model with consideration of new investments). Furthermore, a more rigorous analysis of the expected NPV and risk levels for the efficient portfolios presented in Tables A1 and A9 (existing technologies) and Tables A2 and A10 (existing technologies and new investments) in the Appendix A reveals that the portfolios obtained with FSMAD Model 1 perform slightly better than the corresponding ones obtained with the SMAD Model (compare, for example, P3 in Table A1 with P3 in Table A9, and P6 in Table A1 with P5 in Table A9 for existing technologies; for existing technologies combined with new investments, compare P3 in Table A2 with P3 in Table A10, and P7 in Table A3 with P7 in Table A10, and P8 in Table A3 with P8 in Table A10). It can be seen from Table 1 as well that for the given value of the portfolio return, the risk level in FSMAD Model 1 is slightly smaller. It means that the return value for each technology expressed by the expected value from probability theory (see model (5)) is almost the same as in the case of the trapezoidal membership function from model (8).

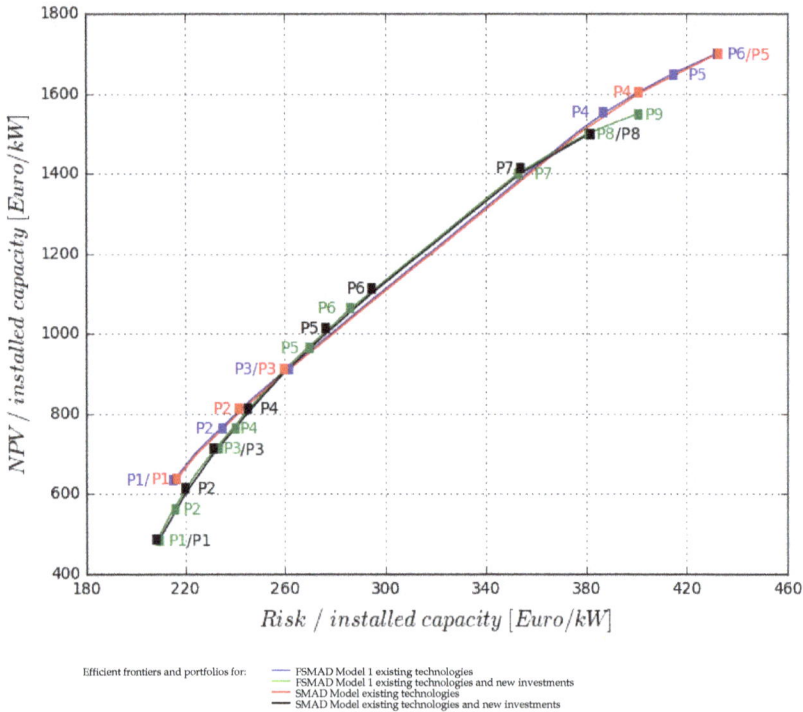

Figure 2. Comparison of efficient frontiers obtained with the FSMAD Model 1 and the SMAD Model.

Figure 3 presents the location of the efficient frontiers for the FSMAD models described in Section 3, together with the SMAD Model and the MV Model (analogously to the other presented models and their efficient frontiers, tables with the technology shares in the efficient power generation mixes for the MV Model are presented in Tables A7 and A8 in the Appendix A). The position of the MV efficient frontier, in contrast to the SMAD Model as well as FSMAD Models 1 and 3, shows the shift in the scale of risk. These two fuzzy models (FSMAD Models 1 and 3) and the semi-mean-absolute deviation model perform better. They achieve a smaller expected risk for the same return level than the other models presented (see also Table 1). Further interesting results observed are the higher gradients of the efficient frontiers for the SMAD Model as well for FSMAD Models 1 and 3, in comparison to the efficient frontiers obtained with the MV Model and FSMAD Model 2, respectively.

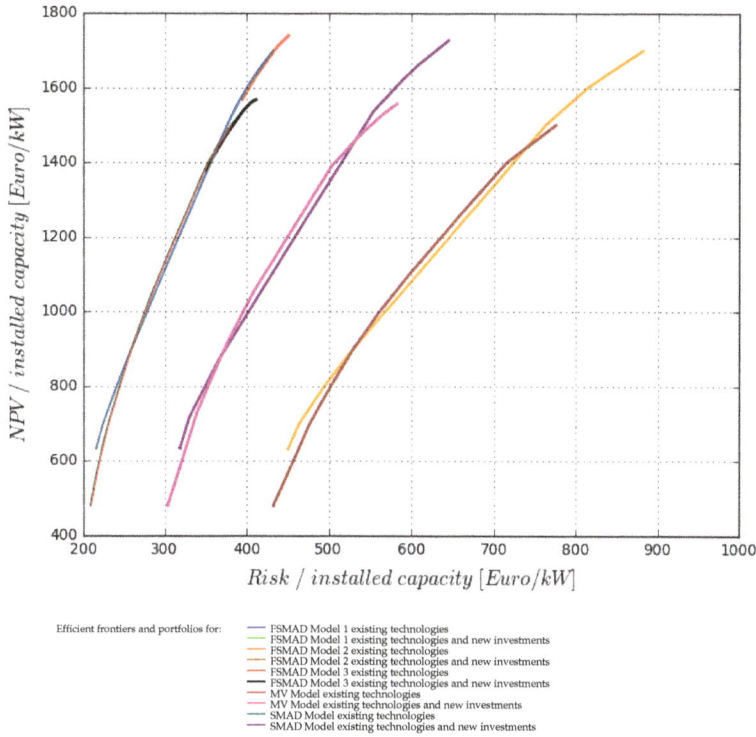

Figure 3. Comparison of the efficient frontiers obtained with all models considered.

Analyzing the comparison of the risk levels for the given portfolio return values (Table 1), it can be noticed that FSMAD Model 2 performs the worst in comparison to all other models. The membership function applied in this model has a trapezoidal form like that in FSMAD Model 1, but a different tolerance interval and left- and right-width parameters (see Section 3.2). The introduction of sufficiency and necessity return and risk levels in FSMAD Model 3 have the effect that the efficient frontier is shorter in comparison to those of the other models (see Figure 1c), and the risk values for smaller portfolio returns are inexistent (see Table 1).

In the following, we compare the results presented in Figure 3 with portfolios obtained when the distribution parameters, especially the variance of the commodities considered in the calculation of the expected returns of the power generation portfolio assets, change. In order to mimick more downside and upside events, we doubled the variance and re-simulated the return values for all considered technologies. Simulation results were applied to the portfolio selection based on the all models presented in the paper. Figure 4 compares the location of efficient frontiers obtained previously—plot (a) (the graphic is similar to Figure 3 but the scale of the risk is changed for better comparison) with efficient frontiers obtained when doubling the variance—plot (b). As can be seen, the efficient frontier for the MV portfolio selection model shifts markedly to the right, implying that portfolios are characterized with higher risk for the same expected return in comparison to the previous situation. In contrast, the efficient portfolios obtained according to FSMAD Model 1 and the SMAD Model shift only slightly to the left. This observation proves the previously mentioned property of the SMAD Model, i.e., that the SMAD as a risk measure is less sensitive to outliers and equivalent to the MV approach. The efficient frontier obtained for FSMAD Model 2 also shifts strongly to the right, which can be a consequence of the membership function applied in this model (i.e. one which uses the quantiles, which is the cause of large variance change, too).

Table 1. Risk values for all models for selected portfolio returns [both in €/kW].

Value of the Portfolio Return	Model				
	FSMAD 1	**FSMAD 2**	**FSMAD 3**	**MV**	**SMAD**
(a) Existing technologies					
633	215.26	450.35	-	318.89	216.33
900	258.79	529.21	-	366.85	259.58
1700	432.02	882.27	450.84	645.13	432.00
(b) Existing technologies and new investments					
479	208.47	432.38	-	304.10	209.46
700	230.58	477.07	-	399.65	231.42
1400	352.95	715.97	357.96	540.34	353.82
1500	380.93	775.99	390.05	564.89	381.65

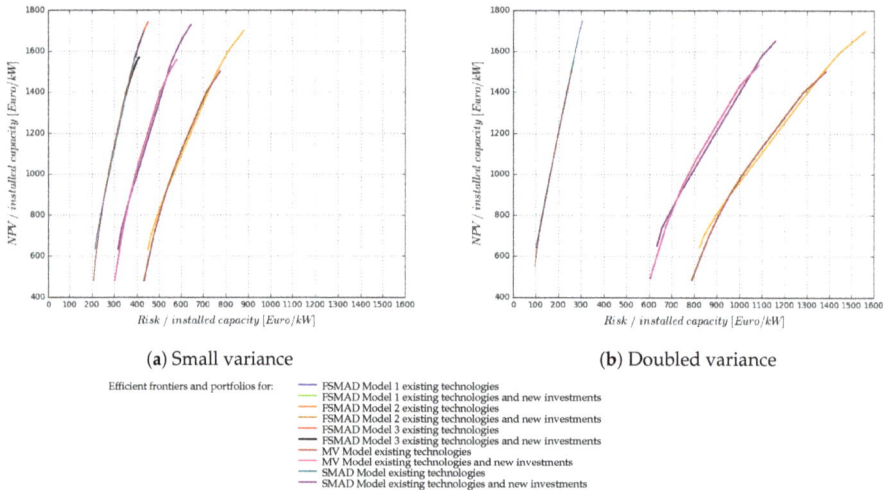

(a) Small variance (b) Doubled variance

Efficient frontiers and portfolios for:
- FSMAD Model 1 existing technologies
- FSMAD Model 1 existing technologies and new investments
- FSMAD Model 2 existing technologies
- FSMAD Model 2 existing technologies and new investments
- FSMAD Model 3 existing technologies
- FSMAD Model 3 existing technologies and new investments
- MV Model existing technologies
- MV Model existing technologies and new investments
- SMAD Model existing technologies
- SMAD Model existing technologies and new investments

Figure 4. Comparison of the efficient frontiers obtained with all models considered.

Table 2 compares the risk levels obtained for the given portfolio return values when the variance is doubled. In the case of FSMAD Model 1 and the SMAD Model, the risk value decreases slightly in comparison to the previous situation presented in Table 1 (compare also the shift of the efficient frontiers in Figure 4). For the FSMAD Model 2 as well as the MV Model, the portfolio risk increases strongly. The most interesting result, however, can be observed for FSMAD Model 3, where the risk value increased most (see Table 2), and where the investor's aspiration levels regarding risk and return were included. As mentioned in Section 3.2, these levels can be defined by the decision-makers and also calculated using mathematical methods, such as those proposed by [41] and used in our case. Hence, the efficient frontier for this portfolio selection model is not presented in Figure 4b.

Table 2. Risk values for all models for selected portfolio returns, doubled variance [both in €/kW].

Value of the Portfolio Return	Model				
	FSMAD 1	FSMAD 2	FSMAD 3	MV	SMAD
(a) Existing technologies					
650	104.14	826.08	-	635.87	104.54
850	137.77	960.56	-	746.05	138.18
1750	307.61	1563.66	8760.03	1159.59	308.59
(b) Existing technologies and new investments					
600	103.21	931.59	-	611.38	102.96
1050	174.99	831.59	-	810.16	173.49
1550	268.18	1384.35	7958.00	1085.83	267.73

5. Conclusions

In this paper, we have presented several alternative portfolio selection models for power generation assets based on fuzzy sets theory and semi-mean-absolute deviation as a risk measure. On the one hand, the use of another risk measure than the standard deviation (as it is used in the standard Markowitz model) was already suggested by Markowitz himself, but the first application only emerged in the aftermath of post-modern portfolio theory. On the other hand, measures such as the semi-variance or semi-absolute deviation, from an investor's point of view, describe the expected losses and thus the part of the risk (the downside risk) that really matters risk. For this reason, the consideration of these measures in decision-making processes seems to be necessary, or even indispensable. In the presented results, the use of the SMAD Model caused a shift of the efficient frontier along the risk axis. More precisely, the efficient portfolios for the same return level have a smaller risk than portfolios obtained with the MV Model. For a decision-maker with risk aversion, such a shift can positively affect the decision-making outcome.

The analysis carried out in this paper illustrates the application possibilities of fuzzy portfolio selection models for power generation assets. Specifically, introducing membership functions for the description of investors' aspiration levels for the expected return and risk (FSMAD Model 3) shows how the knowledge of experts, and an investor's subjective opinions, can be better integrated into the decision-making process. In the cases presented, we have shown that using one of these models affects the size of the set of efficient portfolios (the set is smaller than when using FSMAD, SMAD or MV models). The sparse set of alternatives, which is considered in the decision-making process, can push on and relieve this process. Moreover, in FSMAD Model 3, the decision-maker can help to exactly determine the so-called sufficiency and necessity levels and obtain the optimal solution.

The fuzzy portfolio selection models and the model that uses the semi-mean-absolute deviation as a risk measure presented in this paper illustrated their application for energy utilities, just as they exist for other industries and the financial markets. The complexity of the energy markets, the uncertain environment, the vagueness or some other type of fuzziness, can overall be better captured with fuzzy sets theory. However, the further development of these models especially for the energy sector is required, which calls for more research in this field as well as applications of other alternative risk measures, such as, e.g., the CVaR.

Author Contributions: The authors conceived the models of fuzzy sets theory for portfolio optimization problems in combination with alternative portfolio risk measures. The authors applied the models to power generation assets and wrote the paper together. B.G. programmed all models in Python 2.6.

Funding: The authors gratefully acknowledge the financial support provided by E.ON ERC gGmbH, Project No. 02-01 "Optimization of E.ON's Power Generation with a Special Focus on Renewables".

Conflicts of Interest: The authors declare no conflicts of interest.

Appendix A

Table A1. Efficient portfolios according to FSMAD Model 1 for existing technologies.

Technologies	Efficient Portfolios					
	P1	P2	P3	P4	P5	P6
Biomass	0.07%	0.07%	0.07%	0.07%	0.07%	0.07%
CCGT	9.19%	7.13%	0.00%	0.00%	8.47%	9.19%
CHP	8.53%	8.53%	8.53%	8.53%	8.53%	5.70%
GT gas	3.24%	3.24%	3.24%	3.24%	0.00%	0.00%
GT oil	7.24%	7.24%	7.24%	7.24%	6.50%	0.00%
Hard coal	45.67%	45.67%	44.19%	11.01%	0.00%	8.61%
Hydro	14.95%	14.95%	14.95%	14.95%	14.95%	14.95%
Lignite	8.22%	0.00%	0.00%	1.69%	8.22%	8.22%
Nuclear	2.46%	12.74%	21.35%	52.84%	52.84%	52.84%
Onshore	0.42%	0.42%	0.42%	0.42%	0.42%	0.42%
NPV [€/kW]	632.71	750.00	900.00	1550.00	1650.00	1700.00
Risk [€/kW]	215.26	232.02	258.79	386.84	414.56	432.02

Table A2. Efficient portfolios according to FSMAD Model 1 for existing technologies combined with new investments.

Technologies	Efficient Portfolios								
	P1	P2	P3	P4	P5	P6	P7	P8	P9
Biomass	0.06%	0.06%	0.06%	0.06%	0.06%	0.06%	0.06%	0.06%	0.06%
CCGT	16.36%	16.36%	16.36%	15.33%	5.93%	0.00%	0.00%	7.83%	7.83%
CHP	7.26%	7.26%	7.26%	7.26%	7.26%	7.26%	7.26%	7.26%	0.55%
GT gas	2.76%	2.76%	2.76%	2.76%	2.76%	2.76%	2.76%	2.76%	0.00%
GT oil	6.17%	6.17%	6.17%	6.17%	6.17%	6.17%	6.17%	6.17%	0.00%
Hard coal	48.76%	43.48%	38.88%	38.88%	38.88%	38.88%	21.73%	9.53%	22.74%
Hydro	12.73%	12.73%	12.73%	12.73%	12.73%	12.73%	12.73%	12.73%	12.73%
Lignite	1.79%	7.00%	2.70%	0.00%	0.00%	0.00%	0.92%	7.00%	7.00%
Nuclear	0.00%	0.07%	8.97%	12.69%	22.10%	28.03%	44.98%	44.98%	44.98%
Offshore	3.76%	3.76%	3.76%	3.76%	3.76%	3.76%	3.76%	3.76%	3.76%
Onshore	0.36%	0.36%	0.36%	0.36%	0.36%	0.36%	0.36%	0.36%	0.36%
NPV [€/kW]	479.71	550.00	700.00	750.00	950.00	1050.00	1400.00	1500.00	1550.00
Risk [€/kW]	208.47	214.14	230.58	237.33	266.57	284.00	352.95	380.93	400.58

Table A3. Efficient portfolios according to FSMAD Model 2 for existing technologies.

Technologies	Efficient Portfolios				
	P1	P2	P3	P4	P5
Biomass	0.07%	0.07%	0.07%	0.07%	0.07%
CCGT	9.19%	4.17%	0.00%	0.00%	9.19%
CHP	8.53%	8.53%	8.53%	8.53%	5.70%
GT gas	3.24%	3.24%	3.24%	0.00%	0.00%
GT oil	7.24%	7.24%	7.24%	0.00%	0.00%
Hard coal	45.67%	45.67%	44.19%	9.56%	8.61%
Hydro	14.95%	14.95%	14.95%	14.95%	14.95%
Lignite	8.22%	0.00%	0.00%	6.38%	8.22%
Nuclear	2.46%	15.71%	21.35%	52.84%	52.84%
Onshore	0.42%	0.42%	0.42%	0.42%	0.42%
NPV [€/kW]	632.71	800.00	900.00	1600.00	1700.00
Risk [€/kW]	450.35	494.73	529.21	814.72	882.27

Table A4. Efficient portfolios according to FSMAD Model 2 for existing technologies combined with new investments.

Technologies	Efficient Portfolios							
	P1	P2	P3	P4	P5	P6	P7	P8
Biomass	0.06%	0.06%	0.06%	0.06%	0.06%	0.06%	0.06%	0.06%
CCGT	16.36%	16.36%	16.36%	13.10%	2.96%	0.00%	0.00%	7.26%
CHP	7.26%	7.26%	7.26%	7.26%	7.26%	7.26%	7.26%	7.26%
GT gas	2.76%	2.76%	2.76%	2.76%	2.76%	2.76%	0.00%	0.00%
GT oil	6.17%	6.17%	6.17%	6.17%	6.17%	6.17%	6.17%	0.00%
Hard coal	48.76%	41.33%	38.88%	38.88%	38.88%	36.39%	24.00%	23.23%
Hydro	12.73%	12.73%	12.73%	12.73%	12.73%	12.73%	12.73%	12.73%
Lignite	1.79%	7.00%	2.70%	0.00%	0.00%	0.00%	0.68%	7.00%
Nuclear	0.00%	2.23%	8.97%	14.93%	25.07%	30.52%	44.98%	44.98%
Offshore	3.76%	3.76%	3.76%	3.76%	3.76%	3.76%	3.76%	3.76%
Onshore	0.36%	0.36%	0.36%	0.36%	0.36%	0.36%	0.36%	0.36%
NPV [€/kW]	479.71	600.00	700.00	800.00	1000.00	1100.00	1400.00	1500.00
Risk [€/kW]	432.38	456.76	477.07	502.51	561.46	597.25	715.97	775.99

Table A5. Efficient portfolios according to FSMAD Model 3 for existing technologies.

Technologies	Efficient Portfolios			
	P1	P2	P3	P4
Biomass	0.07%	0.07%	0.07%	0.07%
CCGT	0.00%	0.00%	0.54%	9.19%
CHP	8.53%	8.53%	8.53%	0.05%
GT gas	3.24%	0.00%	0.00%	0.00%
GT oil	7.24%	7.01%	0.00%	0.00%
Hard coal	10.65%	7.95%	14.43%	14.25%
Hydro	14.95%	14.95%	14.95%	14.95%
Lignite	2.05%	8.22%	8.22%	8.22%
Nuclear	52.84%	52.84%	52.84%	52.84%
Onshore	0.42%	0.42%	0.42%	0.42%
NPV [€/kW]	1567.28	1635.69	1676.93	1739.59
Risk [€/kW]	394.61	413.28	426.46	450.84

Table A6. Efficient portfolios according to FSMAD Model 3 for existing technologies combined with new investments.

Technologies	Efficient Portfolios				
	P1	P2	P3	P4	P5
Biomass	0.06%	0.06%	0.06%	0.06%	0.06%
CCGT	0.00%	0.00%	0.00%	2.15%	7.83%
CHP	7.26%	7.26%	7.26%	7.26%	0.00%
GT gas	2.76%	2.76%	0.00%	0.00%	0.00%
GT oil	6.17%	6.17%	5.04%	0.00%	0.00%
Hard coal	23.07%	21.26%	18.81%	21.70%	23.29%
Hydro	12.73%	12.73%	12.73%	12.73%	12.73%
Lignite	0.00%	0.66%	7.00%	7.00%	7.00%
Nuclear	43.84%	44.98%	44.98%	44.98%	44.98%
Offshore	3.76%	3.76%	3.76%	3.76%	3.76%
Onshore	0.36%	0.36%	0.36%	0.36%	0.36%
NPV [€/kW]	1378.85	1408.78	1483.60	1519.44	1566.99
Risk [€/kW]	351.58	357.96	378.59	390.05	411.68

Table A7. Efficient portfolios according to the MV Model for existing technologies.

Technologies	Efficient Portfolios					
	P1	**P2**	**P3**	**P4**	**P5**	**P6**
Biomass	0.07%	0.07%	0.07%	0.07%	0.07%	0.07%
CCGT	9.19%	9.19%	0.00%	0.00%	0.01%	9.19%
CHP	8.53%	8.53%	8.53%	8.53%	8.53%	0.00%
GT gas	3.24%	3.24%	3.24%	0.00%	0.00%	0.00%
GT oil	7.24%	7.24%	7.24%	7.24%	0.00%	0.00%
Hard coal	45.67%	45.67%	45.67%	7.72%	14.96%	14.03%
Hydro	14.95%	14.95%	14.95%	14.95%	14.95%	14.95%
Lignite	8.22%	0.00%	0.00%	8.22%	8.22%	8.22%
Nuclear	2.46%	10.68%	19.87%	52.84%	52.84%	52.84%
Onshore	0.42%	0.42%	0.42%	0.42%	0.42%	0.42%
NPV [€/kW]	634.86	717.45	872.77	1622.72	1663.38	1727.25
Risk [€/kW]	318.89	330.45	366.85	590.29	609.14	645.13

Table A8. Efficient portfolios according to the MV Model for existing technologies combined with new investments.

Technologies	Efficient Portfolios						
	P1	**P2**	**P3**	**P4**	**P5**	**P6**	**P7**
Biomass	0.06%	0.06%	0.06%	0.06%	0.06%	0.06%	0.06%
CCGT	16.36%	16.36%	7.83%	0.00%	0.00%	7.83%	7.83%
CHP	7.26%	7.26%	7.26%	7.26%	7.26%	7.26%	0.00%
GT gas	2.76%	2.76%	2.76%	2.76%	0.00%	0.00%	0.00%
GT oil	6.17%	6.17%	6.17%	6.17%	4.29%	0.00%	0.00%
Hard coal	48.76%	38.88%	38.88%	38.88%	19.56%	16.03%	23.29%
Hydro	12.73%	12.73%	12.73%	12.73%	12.73%	12.73%	12.73%
Lignite	1.79%	0.00%	0.00%	0.00%	7.00%	7.00%	7.00%
Nuclear	0.00%	11.67%	20.20%	28.03%	44.98%	44.98%	44.98%
Offshore	3.76%	3.76%	3.76%	3.76%	3.76%	3.76%	3.76%
Onshore	0.36%	0.36%	0.36%	0.36%	0.36%	0.36%	0.36%
NPV [€/kW]	481.70	729.34	920.59	1052.80	1477.83	1527.36	1556.31
Risk [€/kW]	304.10	339.65	377.32	408.49	540.34	564.89	582.19

Table A9. Efficient portfolios according to the SMAD Model for existing technologies.

Technologies	Efficient Portfolios				
	P1	**P2**	**P3**	**P4**	**P5**
Biomass	0.07%	0.07%	0.07%	0.07%	0.07%
CCGT	9.19%	4.31%	0.00%	0.00%	9.19%
CHP	8.53%	8.53%	8.53%	8.53%	6.3%
GT gas	3.24%	3.24%	3.24%	3.24%	0.00%
GT oil	7.24%	7.24%	7.24%	7.24%	0.00%
Hard coal	45.67%	45.67%	44.32%	6.46%	7.47%
Hydro	14.95%	14.95%	14.95%	14.95%	14.95%
Lignite	8.22%	0.00%	0.00%	6.24%	8.22%
Nuclear	2.46%	15.57%	21.23%	52.84%	52.84%
Onshore	0.42%	0.42%	0.42%	0.42%	0.42%
NPV [€/kW]	634.86	800.00	900.00	1600.00	1700.00
Risk [€/kW]	216.33	241.51	259.58	400.88	432.22

Table A10. Efficient portfolios according to the SMAD Model for existing technologies combined with new investments.

Technologies	Efficient Portfolios							
	P1	P2	P3	P4	P5	P6	P7	P8
Biomass	0.06%	0.06%	0.06%	0.06%	0.06%	0.06%	0.06%	0.06%
CCGT	16.36%	16.36%	16.36%	13.21%	3.13%	0.00%	0.00%	7.83%
CHP	7.26%	7.26%	7.26%	7.26%	7.26%	7.26%	7.26%	7.26%
GT gas	2.76%	2.76%	2.76%	2.76%	2.76%	2.76%	2.76%	2.76%
GT oil	6.17%	6.17%	6.17%	6.17%	6.17%	6.17%	6.17%	4.45%
Hard coal	48.76%	41.42%	38.88%	38.88%	38.88%	36.54%	21.36%	8.82%
Hydro	12.73%	12.73%	12.73%	12.73%	12.73%	12.73%	12.73%	12.73%
Lignite	1.79%	7.00%	2.92%	0.00%	0.00%	0.00%	0.56%	7.00%
Nuclear	0.00%	2.13%	8.75%	14.82%	24.90%	30.37%	44.98%	44.98%
Offshore	3.76%	3.76%	3.76%	3.76%	3.76%	3.76%	3.76%	3.76%
Onshore	0.36%	0.36%	0.36%	0.36%	0.36%	0.36%	0.36%	0.36%
NPV [€/kW]	481.70	600.00	700.00	800.00	1000.00	1100.00	1400.00	1500.00
Risk [€/kW]	209.46	220.07	231.42	245.22	276.17	294.66	353.82	381.65

References

1. Markowitz, H. Portfolio Selection. *J. Financ.* **1952**, *7*, 77–91.
2. Konno, H.; Yamazaki, H. Mean-absolute deviation portfolio optimization model and its application to Tokyo stock market. *Manag. Sci.* **1991**, *37*, 519–531. [CrossRef]
3. Konno, H.; Koshizuka, T. Mean-absolute deviation model. *IIE Trans.* **2005**, *25*, 893–900. [CrossRef]
4. Mansini, R.; Speranza, M. A Heuristic Algorithm for a Portfolio Selection Problem with Minimum Transaction Lots. *Eur. J. Oper. Res.* **1999**, *114*, 219–223. [CrossRef]
5. De Silva, L.; Alem, D.; de Carvalho, F. Portfolio optimization using Mean Absolute Deviation (MAD) and Conditional Value-at-Risk (CVaR). *Production* **2017**, *27*, 1–14. [CrossRef]
6. Liu, S. The mean-absolute deviation portfolio selection problem with interval-valued returns. *J. Comput. Appl. Math.* **2011**, *235*, 4149–4157. [CrossRef]
7. Mansini, R.; Ograczyk, W.; Sparanza, M. Twenty years of linear programming based portfolio optimization. *Eur. J. Oper. Res.* **2014**, *234*, 518–535. [CrossRef]
8. Zimmermann, H. *Fuzzy Set Theory and Its Applications*; Kluwer Academic Publishers: Dordrecht, The Netherlands, 2001.
9. Zadeh, L. Fuzzy sets. *Inf. Control* **1965**, *8*, 338–353. [CrossRef]
10. Zadeh, L. Fuzzy sets as a basis for a theory of possibility. *Fuzzy Sets Syst.* **1978**, *1*, 3–28. [CrossRef]
11. Dubois, D.; Prade, H. *Possibility Theory*; Plenum Press: New York, NY, USA, 1988.
12. Bellman, R.; Zadeh, L. Decision-making in a fuzzy environment. *Manag. Sci.* **1970**, *17*, 141–164. [CrossRef]
13. Fang, Y.; Lai, K.; Wang, S. *Fuzzy Portfolio Optimization: Theory and Methods*; Springer: Berlin/Heidelberg, Germany; New York, NY, USA, 2008.
14. Wang, S.; Zhu, S. On Fuzzy Portfolio Selection Problem. *Fuzzy Optim. Decis. Mak.* **2002**, *1*, 361–377. [CrossRef]
15. Watada, J. Fuzzy Portfolio Selection and its Applications to Decision Making. *Tatra Mt. Math. Publ.* **1997**, *13*, 219–248.
16. Tanaka, H.; Guo, P. Portfolio selection based on upper and lower exponential possibility distribution. *Eur. J. Oper. Res.* **1999**, *114*, 115–126. [CrossRef]
17. Bar-Lev, D.; Katz, S. A Portfolio Approach to Fossil Fuel Procurement in the Electric Utility Industry. *J. Financ.* **1976**, *31*, 933–947. [CrossRef]
18. Madlener, R. Portfolio Optimization of Power Generation Assets. In *Handbook of CO$_2$ in Power Systems*; Zheng, Q., Rebennack, S., Pardalos, P., Pereira, M., Iliadis, N., Eds.; Springer: Berlin/Heidelberg, Germany; New York, NY, USA, 2012; pp. 275–296.
19. Awerbuch, S.; Berger, M. *Applying Portfolio Theory to EU Electricity Planning and Policy-Making*; OECD/IEA: Paris, France, 2003.

20. Roques, F.; Newbery, D.; Nuttall, W. Fuel mix diversification in liberalized electricity markets: A mean-variance portfolio theory approach. *Energy Econ.* **2007**, *30*, 1831–1849. [CrossRef]

21. Krey, B.; Zweifel, P. Efficient and secure power for the United States and Switzerland. In *Analytical Methods for Energy Diversity and Security: Portfolio Optimization in the Energy Sector: A Tribute to the Work of Dr. Shimon Awerbuch*; Bazilian, M., Roques, F., Eds.; Elsevier: New York, NY, USA, 2008; pp. 193–218.

22. Madlener, R.; Wenk, C. *Efficient Investment Portfolios for the Swiss Electricity Supply Sector*; FCN Working Paper No. 2/2008; Institute for Future Energy Consumer Needs and Behavior, Faculty of Business and Economics/E.ON Energy Research Center, RWTH Aachen University: Aachen, Germany, August 2008.

23. Borchert, J.; Schemm, R. Einsatz der Portfoliotheorie im Asset Allokations-Prozess am Beispiel eines fiktiven Anlageraums von Windkraftstandorten. *Z. Energiewirtsch.* **2007**, *31*, 311–322.

24. Glensk, B.; Ganczarek-Gamrot, A.; Trzpiot, G. Portfolio Analysis on Polish Power Exchange and European Energy Exchange. In *Multiple Criteria Decision Making*; Konczak, G., Michnik, J., Nowak, M., Trzaskalik, T., Wachowicz, T., Eds.; UE Katowice: Katowice, Poland, 2013; Volume 8, pp. 18–29.

25. Sharpe, W. Capital Asset Prices: A Theory of Market Equilibrium under Considerations of Risk. *J. Financ.* **1964**, *XIX*, 425–442.

26. Konno, H. Piecewise linear risk function and portfolio optimization. *J. Oper. Res.* **1990**, *33*, 139–156. [CrossRef]

27. Bower, B.; Wentz, P. Portfolio Optimization: MAD vs. Markowitz. *Rose-Hulm. Undergrad. Math. J.* **2005**, *6*, 1–17.

28. Artzner, P.; Delbaen, F.; Eber, J.M.; Heath, D. Coherent measures of risk. *Math. Financ.* **1999**, *9*, 203–228. [CrossRef]

29. Föllmer, H.; Schied, A. *Stochastic Finance: An Introduction in Discrete Time*; Walter de Gruyter: Berlin, Germany, 2002.

30. Pflug, G.; Romisch, W. *Modeling, Measuring and Managing Risk*; World Scientific Publishing: Singapore, 2007.

31. Rockafellar, R.; Uryasev, S.; Zabarankin, M. Generalized deviation in risk analysis. *Financ. Stoch.* **2006**, *10*, 51–74. [CrossRef]

32. Rockafellar, R.; Uryasev, S.; Zabarankin, M. Optimality conditions in portfolio analysis with general deviation measures. *Math. Program. 108* **2006**, *2–3*, 515–540. [CrossRef]

33. Ogryczak, W.; Ruszczynski, A. From stochastic dominance to mean-risk models: Semideviations as risk measures. *Eur. J. Oper. Res.* **1999**, *116*, 33–50. [CrossRef]

34. Mansini, R.; Ograczyk, W.; Sparanza, M. Solvable Models for Portfolio Optimization: A Classification and Computational Comparison. *IMA J. Manag. Math.* **2003**, *14*, 187–220. [CrossRef]

35. Speranza, M. A Heuristic Algorithm for a Portfolio Optimization Model applied to the Milan stock Market. *Comput. Oper. Res.* **1996**, *23*, 433–441. [CrossRef]

36. Carlsson, C.; Fullér, R. On possibilistic mean value and variance of fuzzy numbers. *Fuzzy Sets Syst.* **2001**, *122*, 315–326. [CrossRef]

37. Carlsson, C.; Fullér, R.; Majlender, P. A possibilistic approach to selecting portfolios with highest utility score. *Fuzzy Sets Syst.* **2002**, *131*, 13–21. [CrossRef]

38. Vercher, E.; Bermudez, J.; Segura, J. Fuzzy portfolio optimization under downside risk measures. *Fuzzy Sets Syst.* **2007**, *158*, 769–782. [CrossRef]

39. Huang, X. Portfolio selection with fuzzy returns. *J. Intell. Fuzzy Syst.* **2007**, *18*, 383–390.

40. Chen, G.; Liao, X. A possibilistic Mean Absolute Deviation Portfolio Selection Model. In *Fuzzy Information and Engineering*; Cao, B., Zhang, C., Eds.; Springer: Berlin/Heidelberg, Germany; New York, NY, USA, 2009.

41. Zimmermann, H. Fuzzy programming and linear programming with several objective functions. *Fuzzy Sets Syst.* **1978**, *1*, 45–55. [CrossRef]

42. Vasant, P. Fuzzy production planning and its application to decision making. *J. Intell. Manuf.* **2006**, *17*, 5–12. [CrossRef]

43. Madlener, R.; Glensk, B.; Weber, V. *Fuzzy Portfolio Optimization of Onshore Wind Power Plants*; FCN Working Paper No. 10/2011; Institute for Future Energy Consumer Needs and Behavior, Faculty of Business and Economics/E.ON Energy Research Center, RWTH Aachen University: Aachen, Germany, May 2011.

44. Madlener, R.; Glensk, B.; Raymond, P. *Investigation of E.ON's Power Generation Assets by Using Mean-Variance Portfolio Analysis*; FCN Working Paper No. 12/2009; Institute for Future Energy Consumer Needs and Behavior, Faculty of Business and Economics/E.ON Energy Research Center, RWTH Aachen University: Aachen, Germany, November 2009.

45. Brigham, E.; Ehrhardt, M. *Financial Management: Theory and Practice*; South-Western CENGAGE Learning, Mason, OH, USA, 2008.

46. Madlener, R.; Glensk, B. *Portfolio Impact of New Power Generation Investments of E.ON in Germany, Sweden and the UK*; FCN Working Paper No. 17/2010, Institute for Future Energy Consumer Needs and Behavior, Faculty of Business and Economics/E.ON Energy Research Center, RWTH Aachen University: Aachen, Germany, November 2010.

energies

MDPI

Article

A Graph Theoretic Approach to Optimal Firefighting in Oil Terminals

Nima Khakzad

Faculty of Technology, Policy, and Management, Delft University of Technology, 2628BX Delft, The Netherlands; n.khakzadrostami@tudelft.nl; Tel.: +31-15-278-4709

Received: 8 October 2018; Accepted: 7 November 2018; Published: 9 November 2018

Abstract: Effective firefighting of major fires in fuel storage plants can effectively prevent or delay fire spread (domino effect) and eventually extinguish the fire. If the number of firefighting crew and equipment is sufficient, firefighting will include the suppression of all the burning units and cooling of all the exposed units. However, when available resources are not adequate, fire brigades would need to optimally allocate their resources by answering the question "which burning units to suppress first and which exposed units to cool first?" until more resources become available from nearby industrial plants or residential communities. The present study is an attempt to answer the foregoing question by developing a graph theoretic methodology. It has been demonstrated that suppression and cooling of units with the highest out-closeness index will result in an optimum firefighting strategy. A comparison between the outcomes of the graph theoretic approach and an approach based on influence diagram has shown the efficiency of the graph approach.

Keywords: oil storage plants; domino effect; firefighting; optimization; graph theory; influence diagram

---

1. Introduction

Small fire incidents are a common characteristic of industrial plants containing or processing combustible and flammable substances. Major fires, despite their low-probability, however, are among the most feared types of industrial accidents due to their catastrophic consequences in terms of loss of lives and assets and also a multitude of costly resources needed to control and extinguish them. Examples of major fires include series of fires—known as fire domino effect—at oil storage terminals in the UK in 2005 [1], Puerto Rico in 2009 [2], and in Brazil in 2015 [3], as well as a massive single tank fire in Singapore in 2018 [4].

Due to the importance of major fires and particularly fire domino effects, many works have been devoted to their modelling and risk assessment [5–13]. Engineered fire protection systems such as sprinkler systems are effective in tackling small fires and reducing the probability of small fires escalating to fire domino effects [14], but have reportedly proven ineffective in the event of major fires and already-initiated fire domino effects; this has mainly been due to damage, malfunction, or low performance of engineered fire protection systems when exposed to severe heat of fires [15–18].

Considering the inefficiency of engineered fire protection systems in case of major fires, the intervention of fire brigades to tackle the fires becomes indispensable. Nevertheless, the works devoted to the key role of firefighting in controlling and delaying fire spread have been very few [19,20]. Although available industrial fire protection codes and standards such as NFPA 11 [21], CCPS [22], and API RP-2001 [23] can be used to set baseline firefighting strategies in oil and gas facilities, they do not take into account the facility layout and limited firefighting resources, among other parameters [16,19].

The main goal of firefighting is to extinguish fires before they become large and trigger domino effects. In this regard, firefighters can adopt different strategies, which could be: (i) defensive, where the units exposed to the heat of burning units are cooled using water, (ii) offensive, where attempts are made to suppress burning units, or (iii) mixed, as a combination of the previous two strategies, which is the case in most industrial fire events [19].

If the firefighting resources (personnel, apparatus, etc.) are adequate, a firefighting strategy will include the suppression of all the burning units and the cooling of all the exposed units. However, the main challenge arises when the number of units in danger—both on fire and exposed to fire—exceeds the available firefighting resources, demanding for optimal firefighting strategies to help firefighters answer "which burning units to suppress first and which exposed units to cool first?" in anticipation of more resources becoming available, for instance, from neighboring plants or communities. Considering the two main components of firefighting strategies, the suppression of a burning unit would reduce the emitted heat radiation until the fire is completely extinguished while the cooling of an exposed unit would reduce the amount of heat radiation the unit receives and thus prevents from the spread of the fire to the unit.

The present study is thus aimed at developing a decision support methodology using graph theory for identifying optimal firefighting strategies in the case of insufficient firefighting resources. In this study, we define an optimal firefighting strategy as one which minimizes the likelihood of cascading effects in a spatial network. The application of the methodology is demonstrated via an oil storage plant. Section 2 recaps the basics of graph theory and influence diagram; in Section 3, the methodology is developed, and its application is illustrated in Section 4. The work is concluded in Section 5.

2. Materials

2.1. Graph Theory

A mathematical graph $G = (V, E)$ is a set of nodes (V) and arcs (E). Depending on whether the arcs are directed or undirected, the graph is called directed or undirected, respectively. In a directed graph, a walk from node v_i to v_j is a sequence of connected nodes from the former to the latter where each intermediate node can be traversed several times. A path, however, is a walk in which each intermediate node can be traversed once at most. Accordingly, the geodesic distance between v_i and v_j, denoted as d_{ij} is defined as the length of the shortest path from v_i to v_j. In a directed graph, the distance from v_i to v_j is simply the number of arcs constituting the shortest path from v_i to v_j; if there is no path between v_i and v_j, then $d_{ij} = \infty$.

Based on the concept of geodesic distance, a number of graph metrics can be used to describe the characteristics of the nodes, the arcs, and the graph itself. Among such metrics, closeness centrality scores are very popular in modeling and defining the characteristics of spatial networks and grids [24–26]. The closeness of a node v_i can be defined in two ways: the out-closeness of the node, $C_{out}(v_i)$, as the number of steps needed to reach from the node to every other node of the graph, and the in-closeness, $C_{in}(v_i)$, as the number of steps needed to access v_i from every other node of the graph.

$$C_{out}(v_i) = \frac{1}{\sum_j d_{ij}} \tag{1}$$

$$C_{in}(v_i) = \frac{1}{\sum_j d_{ji}} \tag{2}$$

Based on the nodes centrality scores, the graph's centrality scores can be defined. For instance, the graph's average out-closeness, which is analogous to the graph's average efficiency [27] can be defined as:

$$C_{out}(G) = \frac{1}{N} \sum_{i=1}^{N} C_{out}(v_i) \tag{3}$$

Graph metrics have widely been used to describe and optimize the properties of grid infrastructures and spatial networks such as water distribution networks [28,29], oil storage plants [13], and transportation networks [30].

2.2. Bayesian Network and Influence Diagram

Bayesian network (BN) is a directed graph to represent conditional dependencies among a set of random variables by means of chance nodes and arcs [31,32]. Chance nodes with arcs directed from them are called parent nodes while the ones with arcs directed into them are called child nodes. The nodes with no parents are also called root nodes, whereas the nodes with no children are known as leaf nodes.

Satisfying the Markov condition in BNs, a node is independent of its non-descendant nodes given its parent nodes. As a result, a BN factorizes the joint probability distribution of its nodes as the product of the conditional probability distributions of the variables given their immediate parents (Equation (4)). For the nodes with no parents, i.e., the root nodes, such conditional probabilities are simply replaced by marginal probabilities.

$$P(X_1, X_2, \ldots, X_n) = \prod_{i=1}^{n} P(X_i | pa(X_i)) \tag{4}$$

where $pa(X_i)$ is the parent(s) of node X_i.

A BN can be extended to an influence diagram by adding decision nodes and utility nodes. Each decision node consists of a finite set of decision policies. A decision node should be assigned as the parent of chance nodes whose probability distributions depend on the decision policies. Likewise, the decision node should be the child of chance nodes whose states have to be known to the decision maker before making that specific decision. The utility node U contains utility values (positive or negative) to reflect the preferences of the decision maker regarding the outcome of each decision policy. Among the decision policies d_i ($i = 1, \ldots, m$), one with the highest expected utility (EU) is then selected as the optimal decision d^*:

$$d^* = \underset{d_i}{arg\ max}\ EU(d_i) \quad \text{for } i = 1, \ldots, m \tag{5}$$

3. Methodology

3.1. An Example

Figure 1a depicts the schematics of an oil terminal which consists of six identical gasoline storage tanks. Considering tank fires as the most likely accident scenario, for illustrative purposes, the hypothetical heat radiation intensities emitted from and received by the tanks are presented as a weighted adjacency matrix in Figure 1b, where Q_{ij} is the amount of heat tank T_j receives from a tank fire at tank T_i (kW/m^2). Since all the tanks are atmospheric, the minimum heat radiation to spread the fire from a burning tank to a neighboring tank can be considered as 15 kW/m^2 [33]. It should be noted that having the dimension of the tanks, the separation distances between them, the weather conditions, the type and amount of flammable contents, etc., consequences assessment techniques and software such as ALOHA [34] can be used to calculate accurate amounts of heat radiation. Based on the adjacency matrix, the potential fire spread paths in the terminal can be presented as a directed graph in Figure 1c.

In order to identify optimal firefighting strategies under insufficient resources, we introduce some constraints to the amount and type of the available firefighting resources:

- Due to the limited amount of equipment, the fire brigade would not be able to work on more than two storage tanks at a time, and
- Out of these two storage tanks, due to limited types of equipment, one should be a burning tank and the other an exposed tank. In other words, the firefighters can only afford to suppress a burning tank and to cool another exposed tank simultaneously.
- The aim of the fire brigade would be to prevent/delay the fire spread so that the still-safe storage tanks could be saved. In other words, the suppression of a burning tank, for instance, is performed with the aim of saving the neighboring tanks rather than saving the burning tank itself.

Further, when a storage tank has been exposed to an adjacent burning tank for a while unbeknown to the firefighters, the cooling of the exposed tank would be a more conservative strategy than the suppression of the burning tank [19]; however, in the case of crude oil storage tanks, the suppression of burning tanks should be given priority over the cooling of exposed tanks, due to the imminent risk of boil-over [35].

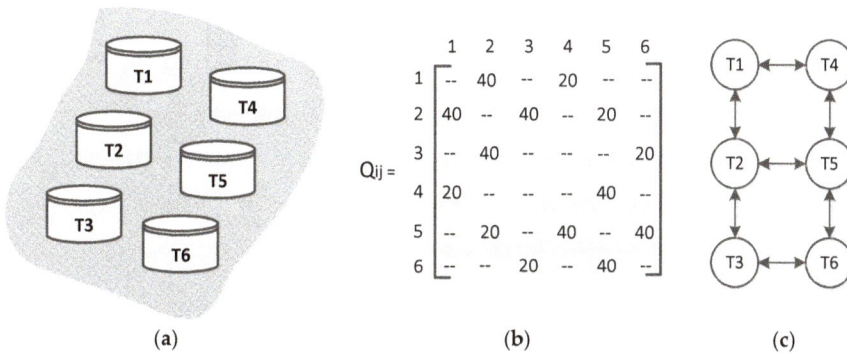

Figure 1. (a) Schematic of a gasoline storage terminal. (b) Mutual heat radiation intensities (kW/m^2) in the event of tank fires. (c) Potential fire spread paths in the terminal as a directed graph.

3.2. Optimal Firefighting Using Graph Theory

Since the identification of firefighting strategies would be based on the observations made by or reported to the firefighters, three fire spread scenarios are considered:

- Scenario 1: fire starts at T1 and spreads to T2;
- Scenario 2: fire starts at T1 and spreads to T2 and T4;
- Scenario 3: two simultaneous fires start at T1 and T5.

Khakzad and Reniers [13] demonstrated that modeling cascading effects in spatial networks as a directed graph, the nodes with a higher out-closeness score (Equation (1)) would result in more extensive failures if selected as the initiating node. Similarly, they illustrated that among spatial networks of different layout but identical number of nodes, the ones with a higher average out-closeness scores (Equation (3)) are more vulnerable to the cascading failures.

3.2.1. Scenario 1: Fire Starts at T1 and Escalates to T2

In the case of fires at T1 and T2, the graph in Figure 1c should be customized to present feasible fire spread paths among the storage tanks. To this end, Figure 2a shows the customized fire spread graph where the storage tanks on fire, T1 and T2, would no longer impact one another, and the double-headed arrows between T1 and T4, T2 and T5, and T2 and T3 have been changed to single-headed arrows directed from the burning tanks to the exposed tanks. Modeling the graph of Figure 2a in igraph package [36], the out-closeness scores of the storage tanks as well as the average out-closeness score of the terminal have been calculated and listed in Table 1 (the 2nd column).

As can be seen from Table 1, between the burning tanks in Figure 2a, the out-closeness of T2 (0.42) is larger than T1 (0.313), implying that T2 would contribute more to the spread of fire through the plant, and thus should be given priority over T1 in being suppressed. Figure 2b shows possible fire escalation patterns in the case T2 is suppressed, and thus there would no longer be any arrows from it to T1, T3, and T5. Recalculating the out-closeness scores of the nodes in Figure 2b (3rd column of Table 1), T4 has the largest out-closeness score among the exposed tanks; thus, it is given priority over the other tanks in being cooled. Figure 2c depicts the graph where T1 is still burning, T2 has been suppressed, and T4 is being cooled. Since T4 is cooled, it would no longer be impacted by T1, which is still burning, and thus the arrow from T1 to T4 should be eliminated. Likewise, since T4 would not get damaged and involved in the fire chain, there is no way that T5 could be damaged and catch fire. This is why the arrow from T4 to T5 should be eliminated, and similarly the arrows from T5 to T6 and T3. As can be noted, the graph's average out-closeness score decreases from 0.48 in Figure 2a to 0.00 in Figure 2c (2nd and 4th columns in Table 1).

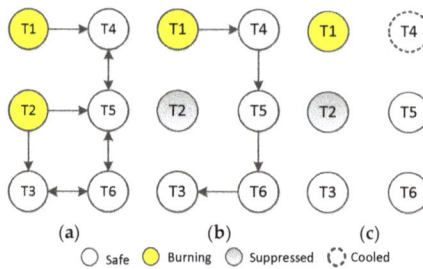

Figure 2. Optimal firefighting strategy for Scenario 1. (**a**) T1 and T2 are on fire. (**b**) T1 is on fire while T2 is suppressed. (**c**) T1 is on fire, T2 is suppressed, and T4 is cooled.

To demonstrate that the suppression of T2 and the cooling of T4 would be the optimal firefighting strategy in Scenario 1, where Figure 3a show the original fire scenario, and Figure 3b depicts a strategy in which T1 is suppressed instead of T2. Given that T1 is suppressed, consider a non-optimal strategy in Figure 4, the out-closeness scores of the tanks are calculated as in Table 1 (the 5th column), indicating T5 with the largest out-closeness score among the exposed tanks and thus the candidate for cooling.

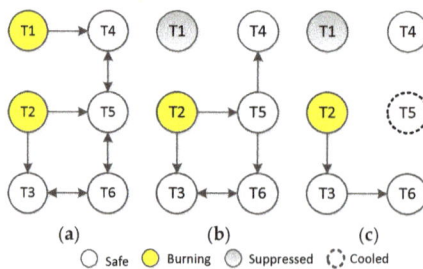

Figure 3. Non-optimal firefighting strategy for Scenario 1. (**a**) T1 and T2 are on fire. (**b**) T1 is suppressed while T2 is on fire. (**c**) T1 is suppressed, T2 is on fire, and T5 is cooled.

Figure 3c depicts the graph where T1 is extinguished, T2 is still on fire, and T5 is cooled. Following this non-optimal strategy, the graph's average out-closeness decreases from 0.48 in Figure 3a to 0.12 in Figure 3c (2nd and 6th columns in Table 1). As can be seen, adopting the former firefighting strategy (Figure 2c) would result in a lower average out-closeness score for the graph than the latter firefighting strategy (Figure 3c), and thus a weaker cascading effect (lower likelihood of fire spread) can be expected [13]. For the sake of clarity, the out-closeness scores of the critical units as well as the average out-closeness score of the plant are depicted with bold numbers in Table 1.

Table 1. Tanks' and terminal's out-closeness scores for the graphs in Figures 2 and 3.

	Out-Closeness Scores				
	Fire Initiation	**Optimal Firefighting**		**Non-Optimal Firefighting**	
Tank	Figures 2a and 3a	Figure 2b	Figure 2c	Figure 3b	Figure 3c
T1	0.31	0.31	0.17	0.17	0.17
T2	**0.42**	0.17	0.17	0.42	0.24
T3	0.28	0.17	0.17	0.20	0.20
T4	0.28	**0.28**	0.17	0.17	0.17
T5	0.31	0.23	0.17	**0.31**	0.17
T6	0.31	0.20	0.17	0.20	0.17
Terminal	**0.48**	0.24	**0.00**	0.24	**0.12**

3.2.2. Scenario 2: Fire Starts at T1 and Spreads to T2 and T4

In this scenario, when firefighters arrive at the scene, the fire has already propagated from T1 to T2 and T4 (Figure 4a). Since T2 has the largest out-closeness score among T1 and T4 (2nd column in Table 2), it is identified as the tank to be extinguished. Given T2 extinguished, and recalculating the out-closeness scores of the tanks in Figure 4b, T5 is identified as the tank with the largest out-closeness score among the other exposed tanks (3rd column in Table 2), and thus chosen for cooling (Figure 4c). Accordingly, the graph's average out-closeness score decreases from 0.24 (2nd column in Table 2) to 0.00 (4th column in Table 2).

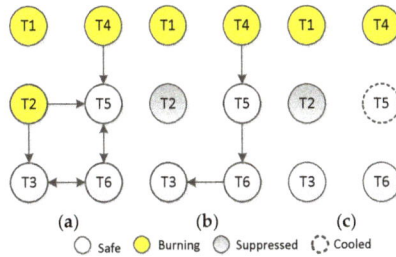

Figure 4. Optimal firefighting strategy for Scenario 2. (**a**) T1, T2, and T4 are all on fire. (**b**) T1 and T4 are on fire while T2 is suppressed. (**c**) T1 and T4 are on fire, T2 is suppressed, and T5 is cooled.

To illustrate the outperformance of the foregoing strategy, the results have been compared with another firefighting strategy where the firefighters decide to extinguish T4 instead of T2 (Figure 5b). Provided that T4 is extinguished, T6 turns out as the exposed tank with the largest out-closeness score (5th column in Table 2); the subsequent cooling of T6 (Figure 5c) would result in a higher average out-closeness score for the graph (0.12) than that of the previous strategy (0.00), indicating the optimality of the former firefighting strategy. For the sake of clarity, the out-closeness scores of the critical units as well as the average out-closeness score of the plant are depicted with bold numbers in Table 2.

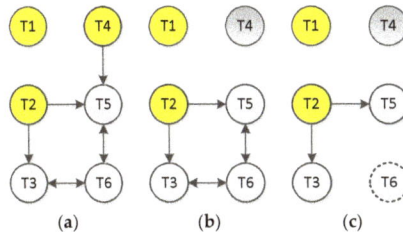

Figure 5. Non-optimal firefighting strategy for Scenario 2. (**a**) T1, T2, and T4 are all on fire. (**b**) T1 and T2 are on fire, and T4 is suppressed. (**c**) T1 and T2 are on fire, T4 is suppressed, and T6 is cooled.

Table 2. Tanks' and terminal's out-closeness scores for the graphs in Figures 4 and 5.

| | Out-Closeness Scores | | | | |
| | Fire Initiation | Optimal Firefighting | | Non-Optimal Firefighting | |
Tank	Figures 4a and 5a	Figure 4b	Figure 4c	Figure 5b	Figure 5c
T1	0.17	0.17	0.17	0.17	0.17
T2	**0.31**	0.17	0.17	0.31	0.25
T3	0.24	0.17	0.17	0.24	0.17
T4	0.28	0.28	0.17	0.17	0.17
T5	0.24	**0.24**	0.17	0.24	0.17
T6	0.25	0.20	0.17	**0.25**	0.17
Terminal	**0.24**	0.16	**0.00**	0.13	**0.12**

3.2.3. Scenario 3: Fire Starts at T1 and T5

In this scenario, when firefighters arrive, there are fires at T1 and T5 as depicted in Figure 6a. Since the out-closeness score of T5 is larger than that of T1 (2nd column in Table 3), suppression of T5 would delay the escalation of fire more effectively. Suppressing T5 in Figure 6b and recalculating the out-closeness scores as reported in the 3rd column in Table 3, T2 turns out as the tank with the largest out-closeness score among the other exposed tanks.

As such, cooling of T2 would better prevent the escalation of fire. Adopting such firefighting strategy, that is, to extinguish T5 and to cool T2, as displayed in Figure 6c, the graph's average out-closeness score decreases from 0.29 in Figure 6a (2nd column in Table 3) to 0.06 in Figure 6c (4th column in Table 3).

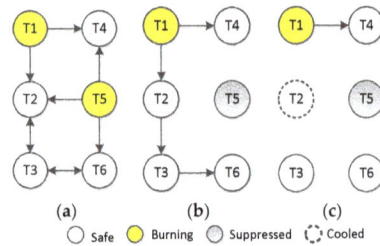

Figure 6. Optimal firefighting strategy for Scenario 3. (**a**) T1 and T5 are on fire. (**b**) T1 is on fire while T5 is suppressed d. (**c**) T1 is on fire, T5 has been suppressed, and T2 is cooled.

If T1 was extinguished instead of T5 (Figure 7b), T3 would be selected as the exposed storage tank with the largest out-closeness score (5th column in Table 3) among the other exposed tanks, thus being chosen for cooling (Figure 7c). This would have led to the graph's average out-closeness score of 0.20 (6th column of Table 3) that is larger than that of Figure 6c, i.e., 0.06 (4th column of Table 3). As such, the firefighting strategy presented in Figure 6c would be preferred than the one depicted in Figure 7c. For the sake of clarity, the out-closeness scores of the critical units as well as the average out-closeness score of the plant are depicted with bold numbers in Table 3.

Table 3. Tanks' and terminal's out-closeness scores for the graphs in Figures 6 and 7.

| | Out-Closeness Scores | | | | |
| | Fire Initiation | Optimal Firefighting | | Non-Optimal Firefighting | |
Tank	Figures 6a and 7a	Figure 6b	Figure 6c	Figure 7b	Figure 7c
T1	0.39	0.39	0.20	0.17	0.17
T2	0.24	**0.24**	0.17	0.24	0.17
T3	0.25	0.20	0.17	**0.25**	0.17
T4	0.17	0.17	0.17	0.17	0.17
T5	**0.46**	0.17	0.17	0.46	0.33
T6	0.24	0.17	0.17	0.24	0.17
Terminal	**0.29**	0.24	**0.06**	0.24	**0.20**

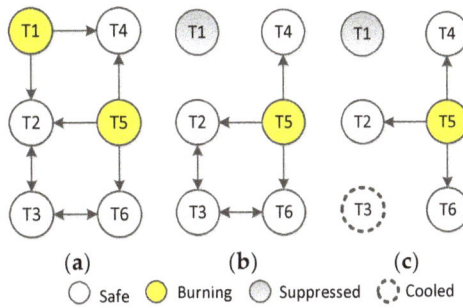

Figure 7. Non-optimal firefighting strategy for Scenario 3. (**a**) T1 and T5 are on fire. (**b**) T1 has been suppressed while T5 is on fire. (**c**) T1 has been suppressed, T5 is on fire, and T3 is cooled.

3.3. Comparison between Graph Theoretic and Influence Diagram Approaches

Khakzad et al. [12] developed a methodology based on BN for domino effect modeling in the chemical and process spatial infrastructures. In their approach, the units are presented as nodes of the BN while the intensity of heat radiation between them are presented as directed arcs. Knowing the intensity of received heat radiation along with the type (e.g., atmospheric or pressurized) and size of exposed units, the probability of fire escalation can be estimated using probit functions [10,33]. Among the exposed units, the one with the highest escalation probability is identified as the secondary unit involved in the fire escalation (domino effect). Following the same approach and considering possible synergistic effects, the tertiary units can be identified.

Figure 8a,b display the BNs for modeling the fire spread in the tank farm given a single fire at T1 (relevant to Scenario 1 and 2) and two separate fires at T1 and T5 (relevant to Scenario 3), respectively. To help distinguish between the graph theoretic approach and the BN approach, the nodes in the BNs have been presented as ellipse. For the sake of clarity, the conditional probability table of node T5 in Figure 8a given its immediate parents T2 and T4 have been reported in Table 4.

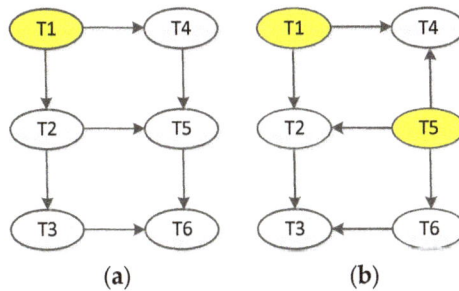

(**a**)   (**b**)

Figure 8. Application of Bayesian network (BN) to modelling fire domino effect in the terminal. (**a**) The domino effect starts from T1. (**b**) The domino effect starts from T1 and T5.

Table 4. Conditional probability table of T5 in Figure 8a.

T2	T4	T5 Fire	T5 Safe
Fire	Fire	P_{24}	$1-P_{24}$
Fire	Safe	P_2	$1-P_2$
Safe	Fire	P_4	$1-P_4$
Safe	Safe	0	1

As previously mentioned, the escalation probabilities in Table 4 can be calculated using probit functions [10,33]. Since the aim of the present study is to identify an optimal firefighting strategy based on relative importance of the storage tanks not their exact escalation probabilities, and because the tanks are of the same type and dimension, we calculate the escalation probabilities using a linear relationship:

$$P_i = 1 - \frac{15}{Q_i} \tag{6}$$

where P_i is the escalation probability of the atmospheric tank exposed to a heat radiation of intensity Q_i (kW/m^2). Further, the numerator 15 in Equation (6) denotes the escalation threshold of atmospheric tanks [33]. Accordingly, the escalation probabilities of tank T5 in Table 4 could be calculated as $P_2 = 1 - 15/20 = 0.25$, $P_4 = 1 - 15/40 = 0.625$, and $P_{24} = 1 - 15/(20 + 40) = 0.75$. It should be noted that the escalation probability given in Equation (6) is merely for illustrative purposes and is not aimed at replacing the probit functions.

The BN in Figure 8a can be extended to influence diagrams in Figure 9a,b in order to identify optimal firefighting strategies in Scenarios 1 and 2. In Figure 9a, for instance, the decision node incorporates eight firefighting strategies (decision alternatives) in form of Ti–Tj (for i = 1, 2 and j = 3, 4, 5, 6), standing for suppression of Ti and cooling of Tj. For example, the decision alternative T1–T3 indicates that the corresponding firefighting strategy involves the suppression of T1 and the cooling of T3. The dashed arcs from T1 and T2 to the decision node implies that the firefighting strategies would be conditioned to the observation of fires at T1 and T2.

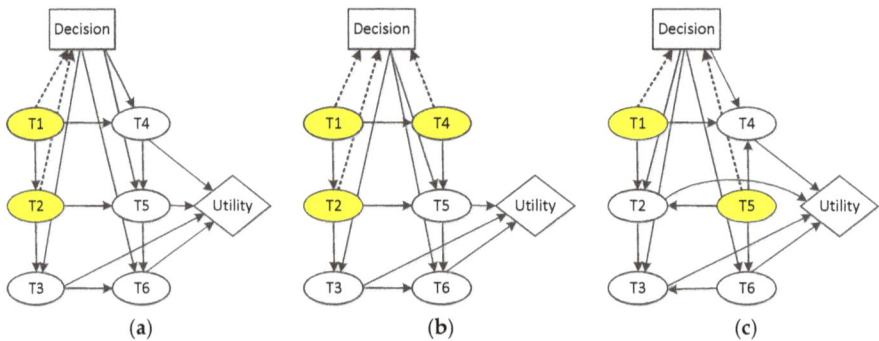

Figure 9. Influence diagrams to identify the optimal firefighting strategies in (**a**) Scenario 1, (**b**) Scenario 2, and (**c**) Scenario 3.

The impact of the decision alternatives on the nodes of the influence diagram can be reflected by making modifications to the conditional probability tables (escalation probabilities) of the exposed tanks which are being influenced by the decision node. Part of the modified conditional probability table of T5 in Figure 9a has been illustrated in Table 5. It should be noted that some entries in Table 5 might seem contradictory at first glance if the entire influence diagram is not taken into consideration.

For example, on the second row of Table 5, the decision T1–T4 denotes the cooling of T4 (i.e., T4 is safe and is being cooled to keep safe) whereas the state of T4 is Fire. Such contradiction can be justified if it is noted that under the same decision alternative the probability of T4 being on fire would be equal to zero (the arc from "Decision" to T4). As such, the escalation probability of T5 (2nd row in Table 5) is only due to the heat radiation received from T2; thus, $P(T5 = Fire \mid d = T1 - T4, T2 = Fire, T4 = Fire) = P_2$. Following the same approach, the BN in Figure 8b can be extended to the influence diagram in Figure 9c to identify the optimal firefighting strategy in the event of Scenario 3.

In the influence diagrams shown in Figure 9a–c, the node "Utility" has been connected to the still-safe storage tanks so that only the further damage caused by the fire spread can be taken into account in the decision making. Since the storage tanks are identical, we have assumed that a damaged storage tank (due to fire spread) is associated with a disutility of −10.0 while the disutility of a safe tank is 0.0. For example, in Figure 9a, U(T3 = Safe, T4 = Fire, T5 = Fire, T6 = Safe) = −20.0.

Table 5. Part of the conditional probability table of T5 in Figure 9a.

Decision	T2	T4	T5	
			Fire	Safe
T1–T3	Fire	Fire	P_{24}	$1-P_{24}$
T1–T4	Fire	Fire	P_2	$1-P_2$
T1–T5	Fire	Fire	0	1
T1–T6	Fire	Fire	P_{24}	$1-P_{24}$
T2–T3	Fire	Fire	P_4	$1-P_4$
T2–T4	Fire	Fire	0	1
T2–T5	Fire	Fire	0	1
T2–T6	Fire	Fire	P_4	$1-P_4$

Implementing the influence diagrams in GeNIe software [37], the expected disutility of firefighting strategies (decision alternatives) are reported in Table 6. In Scenario 1, the optimal decision (attributed to the lowest disutility) would be to suppress T2 and to cool T4 (see the result of the graph theoretic approach in Section 3.2.1). Likewise, in the event of Scenario 2, the suppression of T2 and the cooling of T5 would be the optimal strategy (see the result of the graph theoretic approach in Section 3.2.2) and the vice versa in Scenario 3 (see the result of the graph theoretic approach in Section 3.2.3). As can be seen, the results obtained from the influence diagrams are in complete agreement with those obtained from the graph theoretic approach in the previous sections. For the sake of clarity, the lowest amount of loss (the lowest absolute amount of expected disutility) in each scenario and their corresponding decisions are depicted with bold numbers in Table 6.

Table 6. Expected (Exp) disutility of firefighting strategies for fire scenarios.

Figure 9a: Scenario 1		Figure 9b: Scenario 2		Figure 9c: Scenario 3	
Decision	Exp. Disutility	Decision	Exp. Disutility	Decision	Exp. Disutility
T1–T3	−3.04	T4_T3	−3.04	T1–T2	−8.44
T1–T4	−7.67	T4_T5	−4.64	T1–T3	−9.80
T1–T5	−4.64	T4_T6	−6.00	T1–T4	−7.67
T1–T6	−6.00	T2_T3	−5.24	T1–T6	−6.84
T2–T3	−3.35	**T2_T5**	**0.00**	**T5–T2**	**−2.20**
T2–T4	**0.00**	T2_T6	−3.80	T5–T3	−6.00
T2–T5	−2.20			T5–T4	−5.24
T2–T6	−3.04			T5–T6	−7.44

4. Application of the Methodology

The graph theoretic methodology can effectively be applied to large oil terminals where the number of units could impede the application of influence diagrams. Application of influence diagram to such facilities faces two challenges: (i) for each fire spread scenario which may initiate from a single unit or multiple units a separate influence diagram should be developed, and (ii) due to complicated interactions between the units during fire spread the size of conditional probability tables can grow exponentially and thus becomes intractable.

Figure 10a displays an oil terminal comprising fifteen tanks of gasoline with diameter of $D = 27$ m, height of $H = 15$ m, and capacity of 9000 m^3. Considering tank fires as the most likely fire events, possible fire spread patterns have been presented as the directed graph in Figure 10b.

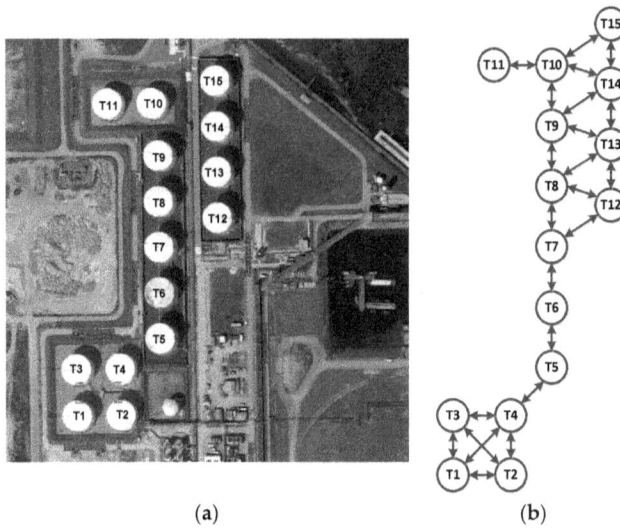

Figure 10. (a) An oil terminal. (b) Possible fire spread scenarios as a directed graph.

Two fire scenarios are considered:

- Scenario A: fire starts at T7 and escalates to T6 and T12, and
- Scenario B: fire starts simultaneously at T4, T10, and T12.

It is also assumed that the plant's firefighting team is equipped with three firefighting trucks, of which two can be used to suppress burning tanks while the third one to cooling one exposed tank. Without replicating the methodology steps, the out-closeness scores calculated sequentially by isolating the suppressed and cooled tanks are reported in Table 7. As can be noted, the following firefighting strategies could be identified as the optimal ones for each fire scenario:

- For Scenario A: suppress T7 and T12, and cool T5,
- For Scenario B: suppress T4 and T12, and cool T8.

Table 7. Identification of optimal firefighting strategies based on the out-closeness scores.

Tank	Scenario A		Scenario B	
	T6, T7 & T12 Are on Fire	T7 & T12 Are Suppressed	T4, T10 & T12 Are on Fire	T4 & T12 Are Suppressed
T1	0.077	0.077	0.077	0.067
T2	0.077	0.077	0.077	0.067
T3	0.077	0.077	0.077	0.067
T4	0.083	0.083	**0.219**	0.067
T5	0.089	**0.089**	0.126	0.067
T6	0.095	0.095	0.136	0.071
T7	**0.114**	0.067	0.144	0.077
T8	0.107	0.067	0.151	**0.152**
T9	0.110	0.067	0.152	0.151
T10	0.109	0.067	0.147	0.147
T11	0.105	0.067	0.136	0.136
T12	**0.117**	0.067	**0.171**	0.067
T13	0.109	0.067	0.149	0.149
T14	0.109	0.067	0.144	0.144
T15	0.107	0.067	0.133	0.133

For the sake of clarity, the out-closeness scores of the critical units are depicted with bold numbers in Table 7. Besides, the modified graphs of Scenarios A and B before and after implementation of the firefighting strategies have been depicted in Figure 11. As can be seen from Figure 11b, in the case of Scenario A, the suppression of T7 and T12 and the cooling of T5 completely prevent from the fire spread in the tank terminal. On the other hand, in the case of Scenario B, the suppression of T4 and T12 and the cooling of T8 do not entirely prevent from but significantly limit the fire spread.

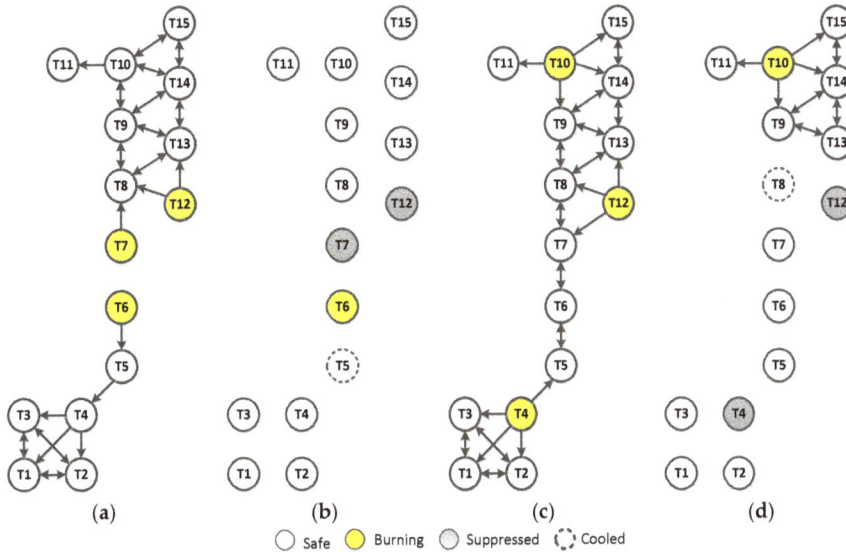

Figure 11. (**a**) Fire spread paths in the case of Scenario A. (**b**) Fire spread paths after the implementation of firefighting in the case of Scenario A. (**c**) Fire spread paths in the case of Scenario B. (**d**) Fire spread paths after the implementation of firefighting in the case of Scenario B.

5. Conclusions

In the present study, we developed methodologies based on graph theory and influence diagrams for optimal firefighting of fires at oil terminals under insufficient firefighting resources. Modeling fire spreads as a directed graph, we demonstrated that suppression of burning units with the highest out-closeness scores would be the most effective fire suppression policy. Removing the suppressed units from the graph, and recalculating the out-closeness scores of the remaining units, it was demonstrated that cooling of exposed units with the highest updated out-closeness scores would be the most effective cooling policy. As such, simultaneous suppressing and cooling of burning and exposed units—as many as the firefighting resources allow—based on sequential calculation of out-closeness scores can be adopted as an optimal firefighting strategy.

In the case of oil and gas facilities which comprise a variety of units of different type and size, the developed influence diagram outperforms the graph theoretic approach by facilitating the incorporation of different escalation probabilities and damage (disutility) values. In the case of large facilities with more or less similar units, however, the graph theoretic approach outdoes the influence diagram. This is mainly because the large number of units can make the development of influence diagrams and identification of conditional probabilities and utility values very time-consuming.

Funding: This research received no external funding. The APC was funded by TUDelft, The Netherlands.

Conflicts of Interest: The author declares no conflict of interest.

References

1. BBC. How the Buncefield Fire Happened. 2010. Available online: http://www.bbc.com/news/uk-10266706 (accessed on 8 October 2018).
2. U.S. Chemical Safety Board (CSB). Caribbean Petroleum Refining Tank Explosion and Fire. 2015. Available online: http://www.csb.gov/caribbean-petroleum-refining-tank-explosion-and-fire (accessed on 8 October 2018).
3. REUTERS. Fuel Tanks on Fire at Storage Facility in SANTOS, Brazil. 2015. Available online: https://www.reuters.com/article/us-brazil-fire-fueltanks/fuel-tanks-on-fire-at-storage-facility-in-santos-brazil-idUSKBN0MT1SU20150402 (accessed on 22 May 2018).
4. REUTERS. Fire Extinguished at Fuel Oil Storage Tank in Singapore. 2018. Available online: https://www.reuters.com/article/us-singapore-energy-fire/fire-extinguished-at-fuel-oil-storage-tank-in-singapore-idUSKBN1GX05N (accessed on 22 May 2018).
5. Vilchez, A.J.; Montiel, H.; Casal, J.; Arnaldos, J. Analytical expressions for the calculation of damage percentage using the probit methodology. *J. Loss Prev. Process. Ind.* **2001**, *14*, 193–197. [CrossRef]
6. Liu, Y. Thermal Buckling of Metal Oil Tanks Subject to an Adjacent Fire. Ph.D. Thesis, University of Edinburgh, Edinburgh, UK, 2011.
7. Mansour, K. Fires in Large Atmospheric Storage Tanks and Their Effect on Adjacent Tanks. Ph.D. Thesis, Loughborough University, Loughborough, UK, 2012.
8. Reniers, G.; Cozzani, V. *Domino Effects in the Process Industries*; Elsevier: Kidlington, UK, 2013.
9. Cozzani, V.; Gubinelli, G.; Antonioni, G.; Spadoni, G.; Zanelli, S. The assessment of risk caused by domino effect in quantitative area risk analysis. *J. Hazard. Mater.* **2005**, *A127*, 14–30. [CrossRef] [PubMed]
10. Landucci, G.; Gubinelli, G.; Antonioni, G.; Cozzani, V. The assessment of the damage probability of storage tanks in domino events. *Accid. Anal. Prev.* **2009**, *41*, 1206–1215. [CrossRef] [PubMed]
11. Abdolhamidzadeh, B.; Abbasi, T.; Rashtchian, D.; Abbasi, S.A. A new method for assessing domino effect in chemical process industry. *J. Hazard. Mater.* **2010**, *182*, 416–426. [CrossRef] [PubMed]
12. Khakzad, N.; Khan, F.; Amyotte, P.; Cozzani, V. Domino effect analysis using Bayesian networks. *Risk Anal.* **2013**, *33*, 292–306. [CrossRef] [PubMed]
13. Khakzad, N.; Reniers, G. Using graph theory to analyze the vulnerability of process plants in the context of cascading effects. *Reliab. Eng. Syst. Saf.* **2015**, *143*, 63–73. [CrossRef]
14. Landucci, G.; Argenti, F.; Tugnoli, A.; Cozzani, V. Quantitative assessment of safety barrier performance in the prevention of domino scenarios triggered by fire. *Reliab. Eng. Syst. Saf.* **2015**, *143*, 30–43. [CrossRef]
15. Nash, P. *Fire Protection of Flammable Liquid Storages by Water Spray and Foam*; Fire Research Station: Borehamwood, UK, 1975.
16. Persson, H.; Lönnermark, A. Tank Fires: Review of Fire Incidents 1951–2003. BRANDFORSK Project 513-021. SP Swedish National Testing and Research Institute, SP Rapport 2004: 14. Available online: https://www.msb.se/Upload/Insats_och_beredskap/Brand_raddning/Oljedepa/Cisternrapport%202004_14.pdf (accessed on 22 May 2018).
17. Lang, X.; Liu, Q.; Gong, H. Study of firefighting systems to extinguish full surface fire of large scale floating roof tanks. *Procedia Eng.* **2011**, *11*, 189–195.
18. Forell, B.; Peschke, J.; Einarsson, S.; Röwekam, M. Technical reliability of active fire protection features—Generic database derived from German nuclear power plants. *Reliab. Eng. Syst. Saf.* **2016**, *145*, 277–286. [CrossRef]
19. D'Amico, M. Risk Based Fire Protection Strategy in Crude Oil Storage Facilities. International Fire Protection Magazine E-Newswire 2015, Issue 64. Available online: https://ifpmag.mdmpublishing.com/risk-based-fire-protection-strategy-in-crude-oil-storage-facilities/ (accessed on 8 October 2018).
20. Khakzad, N. Which fire to extinguish first? A risk-informed approach to emergency response in oil terminals. *Risk Anal.* **2018**, *38*, 1444–1454. [CrossRef] [PubMed]
21. National Fire Protection Association (NFPA). *Standard for Low, Medium, and High-Expansion Foam*; NFPA: Quincy, MA, USA, 2002.
22. Centre for Chemical Process Safety (CCPS). *Guidelines for Fire Protection in Chemical, Petrochemical, and Hydrocarbon Processing Facilities*; Wiley-AIChE: New York, NY, USA, 2003.

23. API RP 2001. Fire Protection in Refineries—Candidate Ballot Draft 8-3-2011. 2011. Available online: http://ballots.api.org/sfp/ballots/docs/RP2001BallotDraft9thEd.pdf (accessed on 8 October 2018).

24. Newman, M. *Networks: An Introduction*; Oxford University Press, Inc.: New York, NY, USA, 2010.

25. Watts, D.J.; Strogatz, S.H. Collective dynamics of 'small-world' networks. *Nature* **1998**, *393*, 440–442. [CrossRef] [PubMed]

26. Albert, R.; Jeong, H.; Barabasi, A.L. Error and attack tolerance of complex networks. *Nature* **2000**, *406*, 378–381. [CrossRef] [PubMed]

27. Latora, V.; Marchiori, M. Efficient behavior of small-world networks. *Phys. Rev. Lett.* **2001**, *87*, 1–4. [CrossRef] [PubMed]

28. Yazdani, A.; Appiah Otoo, R.; Jeffery, P. Resilience enhancing expansion strategies for water distribution systems: A network theory approach. *Environ. Model. Softw.* **2011**, *26*, 1574–1582. [CrossRef]

29. Di Nardo, A.; Di Natale, M. A heuristic design support methodology based on graph theory for district metering of water supply networks. *Eng. Optim.* **2011**, *43*, 193–211. [CrossRef]

30. Huang, H.; Li, K. Train timetable optimization for both a rail line and a network with graph-based approaches. *Eng. Optim.* **2017**, *49*, 2133–2149. [CrossRef]

31. Pearl, J. *Probabilistic Reasoning in Intelligent Systems*; Morgan Kaufmann: San Francisco, CA, USA, 2014.

32. Neapolitan, R. *Learning Bayesian Networks*; Prentice Hall, Inc.: Upper Saddle River, NJ, USA, 2003.

33. Cozzani, V.; Gubinelli, G.; Salzano, E. Escalation thresholds in the assessment of domino accidental events. *J. Hazard. Mater.* **2006**, *129*, 1–21. [CrossRef] [PubMed]

34. ALOHA. US Environmental Protection Agency, National Oceanic and Atmospheric Administration. Available online: http://www.epa.gov/OEM/cameo/aloha.htm (accessed on 8 October 2018).

35. Large Atmospheric Storage Tank Fires (LASTFIRE). LASTFIRE Boilover Lessons, Issue 3, December 2016. Available online: http://www.lastfire.org.uk/refmatpapers.aspx?id=5 (accessed on 8 October 2018).

36. Csardi, G.; Nepusz, T. The igraph software package for complex network research. *Int. J. Complex Syst.* **2006**, *1695*, 1–9.

37. GeNIe. Decision Systems Laboratory, University of Pittsburg. Available online: https://www.bayesfusion.com (accessed on 8 October 2018).

energies

MDPI

Article

A University Building Test Case for Occupancy-Based Building Automation

Siva Swaminathan [1], Ximan Wang [1,2], Bingyu Zhou [3] and Simone Baldi [1,*]

[1] Delft Center for Systems and Control, Delft University of Technology, Mekelweg 2, 2628 CD Delft, The Netherlands; siva1992@gmail.com (S.S.); wangxm614@163.com (X.W.)

[2] System Engineering Research Institute, China State Shipbuilding Corporation, Beijing 100094, China

[3] Siemens AG, Research in Energy and Electronics, Frauenauracher str. 80, 91056 Erlangen, Germany; bingyu.zhou@siemens.com

* Correspondence: s.baldi@tudelft.nl; Tel.: +31-15-2781823

Received: 7 October 2018; Accepted: 9 November 2018; Published: 14 November 2018

Abstract: Heating, ventilation and air-conditioning (HVAC) units in buildings form a system-of-subsystems entity that must be accurately integrated and controlled by the building automation system to ensure the occupants' comfort with reduced energy consumption. As control of HVACs involves a standardized hierarchy of high-level set-point control and low-level Proportional-Integral-Derivative (PID) controls, there is a need for overcoming current control fragmentation without disrupting the standard hierarchy. In this work, we propose a model-based approach to achieve these goals. In particular: the set-point control is based on a predictive HVAC thermal model, and aims at optimizing thermal comfort with reduced energy consumption; the standard low-level PID controllers are auto-tuned based on simulations of the HVAC thermal model, and aims at good tracking of the set points. One benefit of such control structure is that the PID dynamics are included in the predictive optimization: in this way, we are able to account for tracking transients, which are particularly useful if the HVAC is switched on and off depending on occupancy patterns. Experimental and simulation validation via a three-room test case at the Delft University of Technology shows the potential for a high degree of comfort while also reducing energy consumption.

Keywords: heating ventilation and air-conditioning (HVAC); demand side management; occupancy-based control; predicted mean vote (PMV); optimization

1. Introduction

Heating, Ventilation and Air-Conditioning (HVAC) systems, widely used in residential and commercial buildings, are responsible for a large part of the global energy consumption [1]. According to the European Commission's Joint Research Center, Institute for Energy (2009), HVAC systems in the European Union member states were estimated to account for approximately 313 TWh of electricity use in 2007, about 11% of the total 2800 TWh consumed in Europe that year [2]. Energy savings in HVAC systems were therefore identified as a key element to fulfill the target of reducing energy consumption by 20% by 2020. Increased attention has been focused on the reduction of HVAC energy consumption (without violating comfort requirements) [3], via more efficient equipment [4–6], novel approaches to HVAC energy storage [7] or supervisory control techniques [8–10]. A recent literature review of control methods, with an emphasis on the theory and applications of model predictive control for HVAC systems can be found in [11].

Typical HVAC systems are comprised of boilers, air handling units (AHUs), Variable Air Volume (VAV) boxes, radiators, thermal zones, valves, dampers, fans, pumps, pipes and ducts. The primary drawback with the current state of the art is that separate control systems are designed for each

HVAC component, where the design is carried out to ensure that a certain constant reference set point is maintained. Integrating all of these single components require a tedious manual effort by HVAC system installers to tune all these set points: apart from the enormous tuning effort [12], it is difficult to explicitly account for changing conditions, e.g., individual comfort of occupants or their occupancy patterns. Very often, thermal discomfort often leads to constant correction of temperature set-points by the users, causing increased energy consumption [13,14]. Thus, it is necessary to develop a model-based approach with the ability to integrate the human thermal comfort along with various HVAC components. However, as thermal comfort of the users is season-dependent and highly subjective, there exist various attempts to quantify it according to the physical characteristics of both the occupants and their surroundings. Widely-used thermal comfort models are the Adaptive Comfort Model [15] and the Predicted Mean Vote (PMV) [16], where the latter is considered in this work because it is most suited in the absence of natural ventilation. Recent works on occupancy-based building indoor climate control, also touching upon thermal comfort topics, can be found in [17,18].

While there exist many intelligent HVAC control algorithms, they often require the deployment of a completely new control architecture. On the other hand, control architecture for building automation is quite standardized: in particular, most HVAC low-level controllers commissioned in the field today are of Proportional-Integral (PI) or Proportional-Integral-Derivative (PID) type. Therefore, there exists a need to integrate modern controllers with existing PID controllers to ensure that the control objectives are met. Furthermore, current research in Building Management Systems (BMSs) has turned towards Model Predictive Controllers (MPC) for optimal control of building systems, thanks to its capability of handling external disturbances [19], linear and nonlinear models with multiple constraints [20,21]. In view of this situation, in this work, we propose an integrated control structure using an upper MPC layer and lower PID layer. The MPC is based on an integrated HVAC model and generates set-points for the lower layer based on energy and comfort optimization, while the lower level controllers is composed of PI controllers auto-tuned so as to track the reference set-points. One of the main benefits of the integrated control structure is that the PID dynamics can be easily included in the MPC optimization [22]: in this way, we are able to account for tracking transients, which are particularly useful if the HVAC has to be switched on and off depending on occupancy patterns [23]. By doing this, we are able to achieve integration of HVAC and occupants via a PMV index. To the best of our knowledge, the studies available in literature about MPC for office buildings, e.g., [24,25] and references therein, typically neglect such transients. In recent years, there has been a considerable effort in using building energy performance models such as EnergyPlus and TRNSYS [26] not only for simulation and energy consumption purposes, but also for assisting in evaluation of controller design [27,28]. In this work, the proposed control strategy is experimentally validated via an EnergyPlus building energy performance model of a three-room test case at the Delft University of Technology.

This paper will be organized as follows. Section 2 introduces the HVAC test case we consider. Section 3 outlines the optimization problem for both control layers. Section 4 validates the model with real-life data and with EnergyPlus simulations. Section 5 deals with the simulation and results and conclusions are drawn in Section 6. All symbols introduced in the text can be found in Appendix A.

2. Modelling of HVAC Dynamics

We will focus on the cooling test case shown in Figure 1, which models the dynamical interactions between three rooms and one corridor in the Mechanical Engineering faculty of Delft University of Technology (TU Delft). Figure 1 highlights the multi-component interacting structure of the HVAC system, with a chiller driving a cooling coil of an AHU, with the AHU being further connected to a VAV system which supplies fresh air into the rooms and the corridor. Cold water is supplied with a variable-speed pump to the cooling coil, and the fan in the AHU is a variable-speed fan as well. The three rooms have dimensions 16 m^2, 16 m^2 and 20 m^2, with a corridor of 26 m^2. The chiller has capacity of 2 m^3 and maximum energy of 2 kW. The three rooms are subject to a variable occupancy schedule.

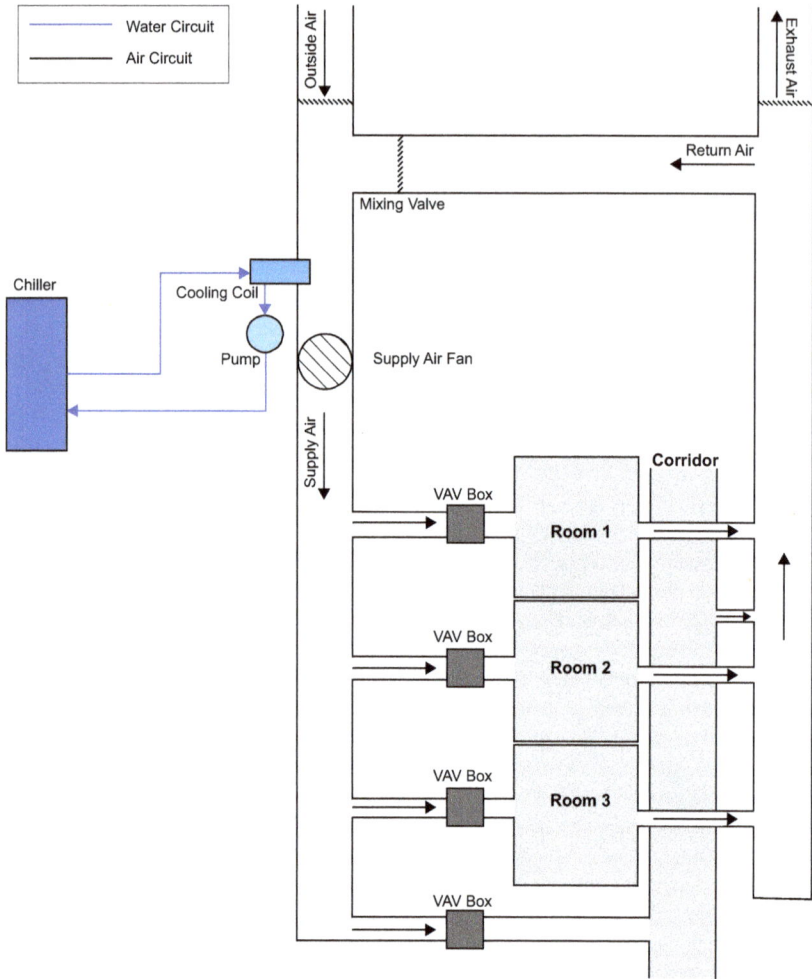

Figure 1. Scheme of the test case Heating, Ventilation and Air-Conditioning system.

The overall integrated model of the HVAC system is a simplified version of the model developed in [29]. The corresponding dynamics are:

$$
\begin{aligned}
\text{[Chiller]} \quad \frac{dT_c}{dt} &= \frac{Q_c}{c_w \rho_w V_c} + \frac{u_w}{\rho_w V_c}(T_{crw} - T_c), \\
\text{[Cooling coil]} \quad \frac{dT_{cc}}{dt} &= \frac{u_w}{\rho_w V_p}(T_c - T_{cc}) + \frac{h_{cc} A_{cc}}{c_w \rho_w V_p}(T_s - T_c), \\
\text{[AHU]} \quad \frac{dT_s}{dt} &= \frac{u_a}{\rho_a V_d}(u_d T_{rm} + (1 - u_d)T_{out} - T_{ma}) \\
&\quad + \frac{h_{cc} A_{cc}}{c_a \rho_a V_d}(T_s - T_{cc}).
\end{aligned}
\tag{1}
$$

At the room level, we quantify the sources and sinks that affect the temperature change in a room, as shown in Figure 2. The balance in room #*i* can be defined via

$$\frac{dT_{rm_i}}{dt} = \underbrace{\frac{u_{rm_i}}{\rho_a V_{rm_i}}(T_s - T_{rm_i})}_{\text{cooling load due to HVAC}} + \underbrace{\frac{h_{wa_i} A_{wa_i}}{c_a \rho_a V_{rm_i}}(T_{wa_i} - T_{rm_i})}_{\text{conduction through walls}}$$

$$+ \underbrace{\frac{h_{wd_i} A_{wd_i}}{c_a \rho_a V_{rm_i}}(T_{wd_i} - T_{rm_i})}_{\text{conduction through windows}} + \underbrace{\frac{q_s}{c_a \rho_a V_{rm_i}}}_{\text{solar radiation}} + \underbrace{q_{int}}_{\text{occupants and equipment}} \quad (2)$$

In Equation (2), T_{wa} and T_{wd} constitute unmeasurable variables. Therefore, the wall and window temperatures are expressed as an affine function of the room and outside air temperature, in line with [30]

$$T_{wa_i} = T_{rm_i} + \frac{T_{out} - T_{rm_i}}{R_{wa_i} A_{wa_i} h_{rm_i}}, \quad (3)$$

where $R_{wa} = \frac{1}{h_{rm_1} A_{wa}} + \frac{l}{k_{wa} A_{wa}} + \frac{1}{h_{out_1} A_{wa}}$ is the combination of conductive and conductive heat transfer coefficients in [K/W]. Similarly, for the window temperature T_{wd}, we have

$$T_{wd_i} = T_{rm_i} + \frac{T_{out} - T_{rm_i}}{R_{wd_i} A_{wd_i} h_{wd_i}}, \quad (4)$$

where $R_{wd} = \frac{1}{h_{rm_2} A_{wd}} + \frac{l}{k_{wd} A_{wd}} + \frac{1}{h_{out_2} A_{wd}}$ is the thermal resistance in [K/W].

A few standard assumptions have been made to develop Equations (1) and (2): air and water are well-mixed and have the same temperature; there is no heat loss through ducts and pipes in the system; thermal conductivity of walls is constant and the heat transfer through it is one-dimensional.

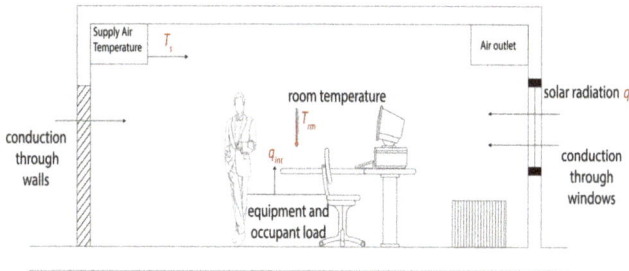

Figure 2. Scheme of the Heating, Ventilation and Air-Conditioning room test case.

In Equations (1) and (2), the multiplication of the control input (flow) by the state (temperature) results in a bilinear system. This continuous-time bilinear model is discretized with $\Delta t = 10$ min using a backward Euler approach, which is well suited for systems with low sampling rates, such as BMSs [31]. The discretized (bilinear) model is then linearized around the point of 24 °C for the room temperature: since the temperature and input range is quite small, this is sufficient for control purposes [32,33]. The resulting discrete-time linear model can be represented in the state-space structure

$$\begin{aligned} x(k+1) &= Ax(k) + Bu(k) + B_d d(k), \\ y(k) &= Cx(k) + Du(k), \end{aligned} \quad (5)$$

where $x = \begin{bmatrix} T_{rm_1} & T_{rm_2} & T_{rm_3} & T_{rm_4} \end{bmatrix}^T \in \mathbb{R}^4$ is the state (comprising the four zone temperatures), $u = \begin{bmatrix} u_{rm_1} & u_{rm_2} & u_{rm_3} & u_{rm_4} \end{bmatrix}^T \in \mathbb{R}^4$ is the input (comprising the four air flow rates), and $d = \begin{bmatrix} T_{out} & q_s & q_{int} \end{bmatrix}^T \in \mathbb{R}^3$ is the disturbance (comprising external temperature, solar radiation and internal gains).

3. Optimization Problem Formulation

The optimization involves: optimization of the low-level PI controls (in order to achieve acceptable tracking of the set points); optimization of the set-points (in order to minimize energy consumption and thermal discomfort). A schematic of the overall control strategy is represented in Figure 3.

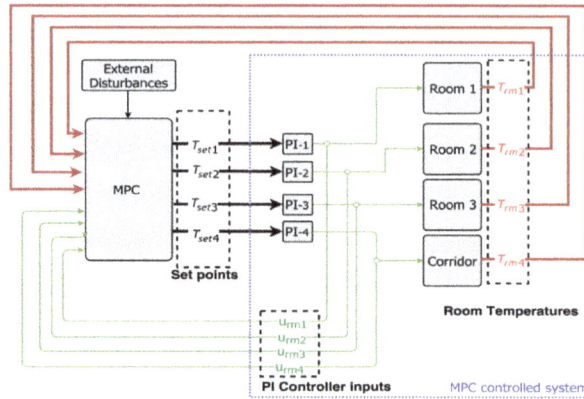

Figure 3. Scheme of proposed control strategy. Note that the feedback of the Proportional-Integral (PI) controllers creates a nested loop with the Model Predictive Control (MPC) layer.

3.1. Optimization for Low-Level Controllers

Four PI controllers, one for each VAV box, are considered: as the purpose of PI control is to achieve tracking with limited energy, we need to quantify the energy consumption of the fan, pump and chiller. With a common duct distributing airflow to all three rooms and the corridor, the total mass airflow u_a blown by the fan is the sum of the individual inlet airflow rates. Therefore, the fan power consumption is [29]

$$Q_f = u_a \Delta P, \tag{6}$$

where ΔP is the total pressure increase in the fan in Pa. The power by the pump is [29]

$$Q_p = \frac{u_w \rho_w g h}{3.6 \cdot 10^6}, \tag{7}$$

where u_w is the pump flow capacity, ρ_w the density of water, g is gravity acceleration and h is the differential head (the term $3.6 \cdot 10^6$ is for the conversion from J to kWh). Finally, the chiller power Q_c is obtained in (1) by calculating the water temperature drop in the cooling coil. All powers are converted into energies after integration over time.

Eight PI gains (four proportional K_p, four integral K_i) are be designed through an offline simulation-based optimization via the MATLAB (Matlab R2016b, The MathWorks, Inc., Natick, Massachusetts, USA) command 'fmincon', to minimize the following cost function:

$$J = \sum_{k=0}^{\tau_f} \left(\sum_{i=1}^{4} (T_{rm_i}(k) - T_{set}(k))^2 \right) + 10^{-2}(Q_f^2 + Q_p^2 + Q_c^2),$$

where T_{set} is the desired zone temperature and τ_f represents the total duration of the simulation (in this case, 24 h). The cost J formalizes the objective to track the desired set points while minimizing energy consumption: the weight 10^{-2} was chosen as a trade-off between these goals. The optimized PI gains for each VAV box are given in Table 1. It must be noted that, because rooms 1 and 2 have identical size and layout, we have imposed the same PI gains: however, even without such imposition, the result of the optimization was having such gains very close to each other.

Table 1. Auto-tuned PI gains after optimization.

Gain	VAV Room 1	VAV Room 2	VAV Room 3	VAV Corridor
K_p	7.63	7.63	8.10	6.92
K_i	0.66	0.66	0.70	0.65

3.2. Optimization for Set-Point Control

3.2.1. Thermal Comfort

The sense of thermal comfort of a human is a highly subjective sensation which could be attributed to various factors such as general health, geographical upbringing and general physical composition. Fanger proposed to quantify such factors and created a predictive model for whole body thermal comfort via the PMV index [34]. The PMV index is now standardized in the American Society of Heating, Refrigerating and Air-Conditioning Engineers (ASHRAE) thermal sensation scale [16]: this thermal scale runs from Cold (−3) to Hot (+3) where 0 indicates maximum user comfort.

The equation for Predicted Mean Vote (PMV) index is

$$\text{PMV} = [0.303e^{-0.036M} + 0.028]L, \tag{8}$$

where L is the thermal load, defined as the difference of metabolic heat generation and the calculated heat loss from the body to the actual environmental conditions, assuming optimal comfort conditions:

$$\begin{aligned} L &= M - W - 3.96 \times 10^{-8} f_{cl}[(t_{cl} + 273)^4 - (t_r + 273)^4] \\ &\quad - f_{cl}h_c(t_{cl} - T_{rm}) - 3.05[5.73 - 0.007(M - W) - \rho_a] \\ &\quad + 0.42[(M - W) - 58.15] - 0.0173M(5.87 - \rho_a) \\ &\quad - 0.0014M(34 - T_{rm}), \end{aligned} \tag{9}$$

where f_{cl} is the clothing factor, h_c is the convective heat transfer coefficient, M is the metabolic rate [W/m²], ρ_a is the vapor pressure of air [kPa], t_{rm} is the room air temperature, t_{cl} is the temperature of the clothing surface [°C], t_r is the mean radiant temperature [°C], and W is the external work (taken as 0 for office conditions).

The mean radiant temperature is a difficult quantity to measure, since it involves measurement of the wall envelope and window temperature [35]. It is also a highly nonlinear function, which can be computationally expensive when included in the cost of the optimization. To overcome this, Rohles [36] proposed an adapted model of the PMV which expresses the thermal sensation as a function of parameters easily sampled in an office environment, such as air temperature and relative humidity. The boundary conditions of the modified PMV index were: clothing insulation level $I_{cl} = 0.6$ clo, metabolic rate $M = 70$ W/m^2, and air velocity $v_a = 0.2$ m/s. With these approximations, the PMV equation from (8) can be expressed as a function of Temperature T_{rm} and water vapour pressure ρ_a, and given by

$$PMV_{rm} = aT_{rm} + b\rho_a - c,\tag{10}$$

where a, b and c are Rohles' experimental coefficients, and are dependent on the gender of the occupants. For a male occupant, $a = 0.212$, $b = 0.293$, $c = 5.949$ and for a female it is $a = 0.275$, $b = 0.255$, $c = 8.62$. The simplified PMV index (10) is used in the predictive optimization.

3.2.2. Model Predictive Controller

To account for tracking transients, we augment the system state x with $\bar{x} = \begin{bmatrix} x^T & x_c^T \end{bmatrix}^T$, where $x_c(k) \in \mathbb{R}^4$ represents the PI controller states. Substituting for input $u_c(k)$ in (5), we have

$$\begin{aligned}
\bar{x}(k+1) &= A_{in}\bar{x}(k) + B_{in}e(k),\\
u_c(k) &= C_{in_u}\bar{x}(k) + D_{in_u}e(k),\\
y(k) &= C_{in_y}\bar{x}(k) + D_{in_y}e(k),
\end{aligned}\tag{11}$$

with $u_c(k) \in \mathbb{R}^4$ being the PI controller inputs and $e(k) \in \mathbb{R}^4$ being the error vector

$$A_{in} = \begin{bmatrix} BC_c & A \\ A_c & 0 \end{bmatrix} \quad B_{in} = \begin{bmatrix} BD_c \\ B_c \end{bmatrix},$$

$$C_{in_u} = \begin{bmatrix} 0 & C_c \end{bmatrix} \quad D_{in_u} = D_c,$$

$$C_{in_y} = \begin{bmatrix} C & DC_c \end{bmatrix} \quad D_{in_y} = DD_c.$$

Substituting back for $e(k)$, we get the overall closed-loop equations with PI controllers

$$\begin{aligned}
A_{out} &= A_{in} - B_{in}(I + D_{in_y})^{-1}C_{in_y},\\
B_{out} &= B_{in} - B_{in}(I + D_{in_y})^{-1}D_{in_y},\\
C_{out_u} &= C_{in_u} - D_{in_y}(I + D_{in_y})^{-1}C_{in_u},
\end{aligned}$$

$$\begin{aligned}
D_{out_u} &= D_{in_u} - D_{in_u}(I + D_{in_y})^{-1}D_{in_u},\\
C_{out_y} &= (I + D_{in_y})^{-1}C_{in_y},\\
D_{out_y} &= (I + D_{in_y})^{-1}D_{in_y},
\end{aligned}\tag{12}$$

which finally gives us the complete state space dynamics of the closed-loop system (the blue dashed box in Figure 3)

$$\begin{aligned}
\bar{x}(k+1) &= A_{out}\bar{x}(k) + B_{out}w(k) + B_d d(k),\\
u_c(k) &= C_{out_u}\bar{x}(k) + D_{out_u}w(k),\\
y(k) &= C_{out_y}\bar{x}(k) + D_{out_y}w(k),
\end{aligned}\tag{13}$$

where $w(k)$ is a vector of set-point temperatures with $w = [T_{set_1} \; T_{set_2} \; T_{set_3} \; T_{set_4}]^T$.

Using (10) and the closed loop state space derived in (11), we formulate the optimization for the MPC as follows:

$$\underset{\tilde{w}(k)}{\text{minimize}} \sum_{k=0}^{N_p-1} \Bigg(\underbrace{||K_u \Delta u||_1}_{\text{energy minimization}} +$$

$$\underbrace{||K_{pmv}(PMV_{rm}(k) - s(k))||_1}_{\text{comfort maximization}} \Bigg),$$

subject to: (11)

$$-0.2 \leq s \leq 0.2, \qquad 0.01 \leq u \leq 2,$$
$$18 \leq y \leq 26, \qquad -0.5 \leq \Delta u \leq 0.5,$$

where $\tilde{w}$ indicates the sequence of set points w along the prediction horizon, and s is a vector of slack variables for the PMV. We set a prediction horizon of $N_p = 10$ for the optimization problem.

4. Validation

To test the real-world feasibility of this approach, we model the building facility at TU Delft using EnergyPlus (EnergyPlus 7.0.0, Department of Energy's (DOE) Building Technologies Office (BTO), Washington, DC, USA) [26], as shown in Figure 4. EnergyPlus is a simulation program that allows simulation the energy consumption for HVAC loads as well as water usage within buildings. Upon constructing an EnergyPlus model of the building, this model was compared with the actual energy usage collected by the Building Management System of the faculty, which is MetaSys (Metasys 9.0, Johnson Controls Inc., Cork, Ireland) by Johnson Controls (a sample interface is shown in Figure 5). Figure 6 shows the experimental simulation of the EnergyPlus model to compare the daily heating demand of the actual and EnergyPlus building.

For the validation, the following simplifications were made. The chiller power was approximated by the electricity consumption of the entire building by scaling it proportionally to the ratio between the volume of the building and the volume of the test rooms and corridor; the damper proportion was kept constant according to information received from the facility management (70:30% mixing of fresh and return air). Finally, together with the EnergyPlus model, a MATLAB model of the building was constructed from (1) and (2) taking into account interactions among rooms (the equations for the entire model cannot be shown due to limited space): all the parameters in the MATLAB model have been derived based on physical properties (density, thermal capacitance, convective heat coefficients). The temperature for the chiller and the cooling coil have been selected as suggested by the facility management. The parameters were further tuned using a system identification procedure, as proposed by [30].

Figure 4. Model of Tower C at TU Delft developed using DesignBuilder and simulated in EnergyPlus.

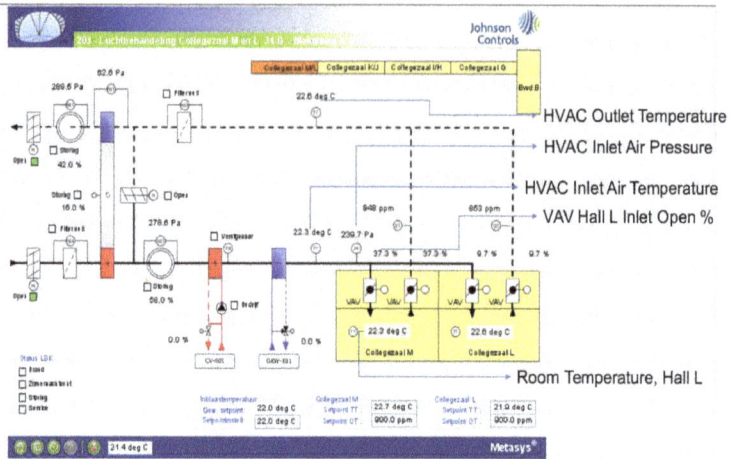

Figure 5. Screenshot of the Building Management System interface of Tower C.

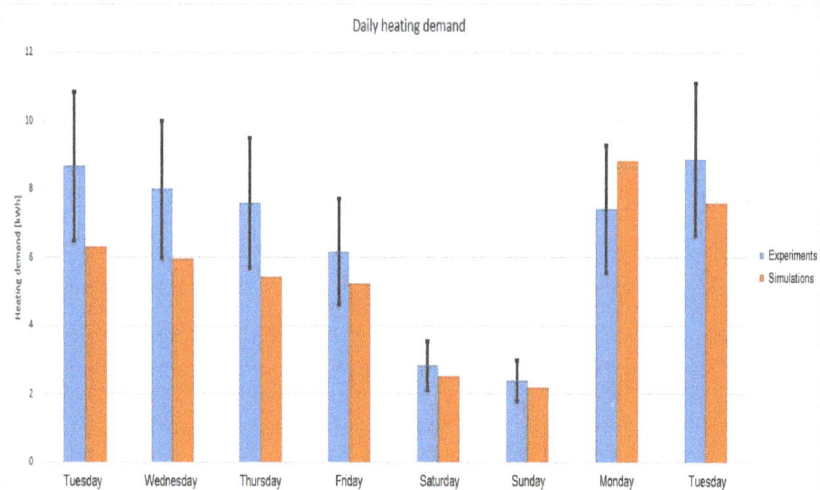

Figure 6. Validation of daily heating demand in EnergyPlus.

5. Simulations

The proposed MPC + Autotuned PI strategy is simulated in MATLAB and interfaced with EnergyPlus. To highlight energy savings, this strategy is compared with a baseline control that tracks a constant set point of 24 °C. Simulations are run for a span of 24 h, with weather profile taken from 19 June 2017, as shown in Figure 7. Please note that the strategy we used has been taken from the actual strategy used in the University building (constant set point and constant 70:30% mixing of fresh and return air). We agree that smarter strategies are in general possible and would lead to different numerical results.

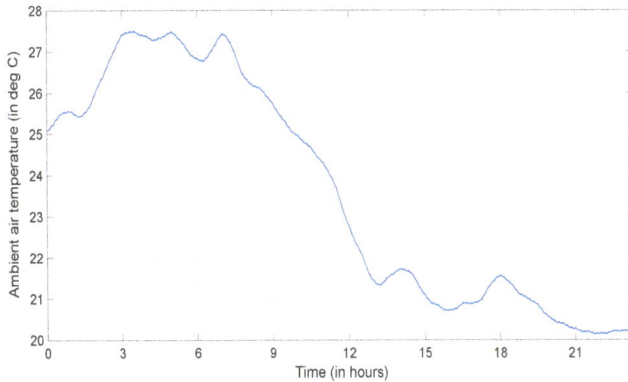

Figure 7. Ambient weather temperature for 19 June 2017.

Figures 8 and 9 show the temperature tracking and PMV profile for two rooms (the other room and the corridor have a similar behavior to the one shown here). The error in set-point tracking is less than ±0.5 °C, which is acceptable due to quantized measurements provided by the sensors simulated in EnergyPlus. When occupants are present in a room, it can be seen that PMV is mostly maintained within ±0.2, which is within the prescribed ASHRAE limits of 0.5. It can also be noted from Figure 10 that the effort is to maintain comfort while having minimal supply air whenever possible.

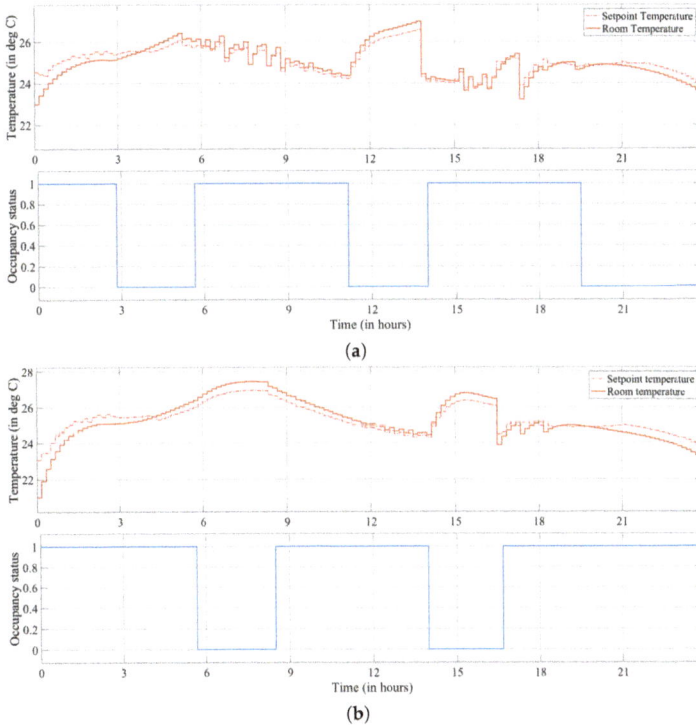

Figure 8. (**a**) temperature profile and set-point tracking, Room 1; (**b**) temperature profile and set-point tracking, Room 2.

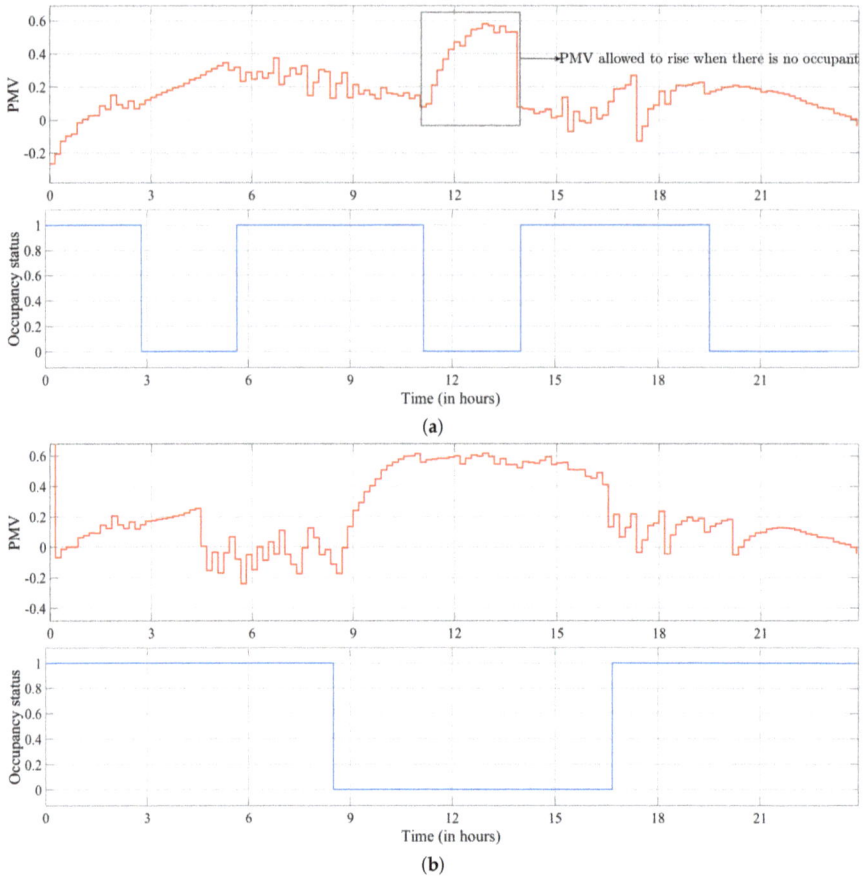

Figure 9. (**a**) evolution of PMV vs. Occupancy, Room 1; (**b**) evolution of PMV vs. Occupancy, Room 3.

The most interesting behavior, which justifies the occupancy-based effort of this work occurs when occupants are not present in a room: in this case, the PMV is allowed to increase (note that the supply air is zero, as shown in Figure 10). Basically, when no people are inside a room, the temperature evolution in the room/corridor is mainly due to conduction through the walls and windows.

In this work, it has been assumed that the occupancy schedule can be forecast. In principle, such forecasting is possible based on the schedule of the lectures: in fact, at TU Delft, the lecture rooms are open during lecture times and closed otherwise. We acknowledge that, in more general settings, such forecasting may be not trivial, c.f. the excellent survey [37]. It can be noted that, because the optimization is based on minimization of PMV, a pre-cooling action is automatically implemented to allow people to find a good climate when they are back. In fact, Figure 10) reveals that around half an hour before people arrive the air flow is turned on again (the other rooms exhibit a similar behavior).

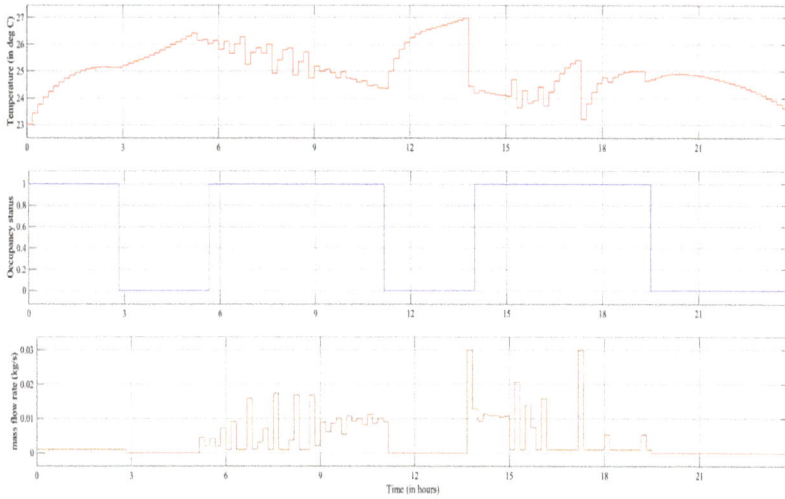

Figure 10. Input air flow, Room 1 (notice the pre-cooling action).

Table 2 shows the comparison of power consumption for the variable supply fan for the baseline PI strategy with the PI with MPC. We notice that, while the optimization only accounted for reduction in fan power, the pump at the chiller side also had a reduced load due to lowered cooling demand. Therefore, there was a significant reduction in the power consumed by the pump as well. Overall, almost 40% reduction in energy consumption of the fan is achieved. Please note that the energy consumption of the fan can be derived from the mass flow rate trend, and it is therefore not reported to avoid including extra figures. In addition, we noticed that the pump and chiller also had a reduced energy consumption due to lowered cooling demand.

Table 2. Air pushed in the rooms in [kg] and total energy consumption in [kW] for a simulation of one day.

Controller	Total Airflow (kg)	Consumption (kW)
Baseline PI	1877.1	12.77
MPC + optimized PI	1132.5 (−39.7%)	7.70 (−39.7%)

6. Conclusions and Future Work

This paper presented an integrated framework to model the set-point and low-level control of a multi-component HVAC system. Starting from a physics-based modelling, a system-of-subsystems dynamic model was used to design a set of strategies that integrate set-point predictive control and low-level PI control. One of the advantages of the proposed control strategy is that, by embedding the PI dynamics in the predictive structure, we are able to account for tracking transients, which are particularly useful if the HVAC is switched on and off depending on occupancy patterns. In addition, we do not disrupt the standard hierarchy of controllers typical of building automation systems. The lower level controllers are auto-tuned so as to track the set points with limited energy consumption; the set points generated by the higher level controller were generated in such a way that comfort was maximized and overall system energy minimized. Energy savings of around 40% have been reported, by using a three-room test case at the Delft University of Technology validated in EnergyPlus using real-life data. Future work will include extension of this design principle so as to consider a stochastic MPC with chance constraints to increase flexibility of the strategy to varying weather and heat load conditions.

Author Contributions: Individual contributions: Conceptualization, S.B.; Methodology, S.S. and S.B.; Software, S.S.; Validation, S.S., X.W. and B.Z.; Formal Analysis, S.B.; Investigation, S.S.; Resources, S.S., X.W., B.Z. and S.S.; Data Curation, S.B.; Writing—Original Draft Preparation, S.S.; Writing—Review and Editing, X.W., B.Z. and S.B.; Visualization, S.S.; Supervision, S.B.; Project Administration, S.B.; Funding Acquisition, S.B.

Funding: This research has been partially funded by the Dutch Central Government Real Estate Agency (Rijksvastgoedbedrijf) under the program Green Technologies 3.0, and by the DCSC department under the Beleidsruimte funding for Distributed Intelligent Climate Control in the DCSC department.

Acknowledgments: The authors gratefully acknowledge the personnel of the TU Delft: Campus and Real Estate (CRE) for providing access to the indoor data.

Conflicts of Interest: The authors declare no conflict of interest.

Appendix A. Nomenclature

Table A1. List of symbols used.

Name	Description
T	Temperature (°C)
Q	Input Power (kW)
u	Control input
c	Specific heat capacity (kJ/kg·K)
ρ	Density (kg/m^3)
V	Volume (m^3)
h	Heat transfer coefficient (W/m^2·K)
A	Area (m^2)
q	Load due to external sources
M	Metabolic rate (W/m^2)
W	Rate of external work (=0 for office conditions)

Subscripts	Description
c	Chiller
cc	Cooling coil
rm	Room
s	Supply air
a	air
w	water
cl	clothing
r	radiant surface

References

1. Pérez-Lombard, L.; Ortiz, J.; Pout, C. A review on buildings energy consumption information. *Energy Build.* 2008, *40*, 394–398. [CrossRef]
2. Knight, I. Assessing Electrical Energy Use in HVAC Systems. 2012. Available online: http://www.rehva.eu/fileadmin/hvac-dictio/01-2012/assessing-electrical-energy-use-in-hvac-systems_rj1201.pdf (accessed on 15 August 2018).
3. Wemhoff, A. Calibration of HVAC equipment PID coefficients for energy conservation. *Energy Build.* **2012**, *45*, 60–66. [CrossRef]
4. Schicktanz, M.; Nunez, T. Modelling of an adsorption chiller for dynamic system simulation. *Int. J. Refrig.* 2009, *32*, 588–595. [CrossRef]
5. Teitel, M.; Levi, A.; Zhao, Y.; Barak, M.; Bar-lev, E.; Shmuel, D. Energy saving in agricultural buildings through fan motor control by variable frequency drives. *Energy Build.* **2008**, *40*, 953–960. [CrossRef]
6. Koh, J.; Zhai, J.Z.; Rivas, J.A. Comparative energy analysis of VRF and VAV systems under cooling mode. In Proceedings of the ASME 3rd International Conference on Energy Sustainability, ES2009, San Francisco, CA, USA, 19–23 July 2009; Volume 1, pp. 411–418.
7. Henze, G.P.; Florita, A.R.; Brandemuehl, M.J.; Felsmann, C.; Cheng, H. Advances in near-optimal control of passive building thermal storage. *J. Sol. Energy Eng.* **2010**, *132*, 021009. [CrossRef]

8. Wang, S.; Ma, Z. Supervisory and optimal control of building HVAC systems: A review. *HVAC R Res.* **2008**, *14*, 3–32. [CrossRef]
9. Ma, Z.; Wang, S.; Xu, X.; Xiao, F. A supervisory control strategy for building cooling water systems for practical and real time applications. *Energy Convers. Manag.* **2008**, *49*, 2324–2336. [CrossRef]
10. Nassif, N.; Moujaes, S. A cost-effective operating strategy to reduce energy consumption in a hvac system. *Int. J. Energy Res.* **2008**, *32*, 543–558. [CrossRef]
11. Afram, A.; Janabi-Sharifi, F. Theory and applications of HVAC control systems—A review of model predictive control (MPC). *Build. Environ.* **2014**, *72*, 343–355. [CrossRef]
12. Kontes, G.D.; Giannakis, G.I.; Kosmatopoulos, E.B.; Rovas, D.V. Adaptive-fine tuning of building energy management systems using co-simulation. In Proceedings of the 2012 IEEE International Conference on Control Applications, Dubrovnik, Croatia, 3–5 October 2012; pp. 1664–1669.
13. Endel, P.; Holub, O.; Berka, J. Adaptive quantile estimation in performance monitoring of building automation systems. In Proceedings of the 2016 European Control Conference (ECC), Aalborg, Denmark, 29 June–1 July 2016; pp. 1189–1194.
14. Giannakis, G.I.; Kontes, G.D.; Kosmatopoulos, E.B.; Rovas, D.V. A model-assisted adaptive controller fine-tuning methodology for efficient energy use in buildings. In Proceedings of the 2011 19th Mediterranean Conference on Control Automation (MED), Corfu, Greece, 20–23 June 2011; pp. 49–54.
15. Brager, G.S.; De Dear, R. *Climate, Comfort, & Natural Ventilation: A New Adaptive Comfort Standard for ASHRAE Standard 55*; American Society of Heating, Refrigerating and air-Conditioning Engineers: Atlanta, GA, USA, 2001.
16. ASHRAE. *ANSI/ASHRAE Standard 55-2010: Thermal Environmental Conditions for Human Occupancy*; American Society of Heating, Refrigerating and air-Conditioning Engineers: Atlanta, GA, USA, 2010.
17. Mirakhorli, A.; Dong, B. Occupancy behavior based model predictive control for building indoor climate—A critical review. *Energy Build.* **2016**, *129*, 499–513. [CrossRef]
18. Baldi, S.; Korkas, C.D.; Lv, M.; Kosmatopoulos, E.B. Automating occupant-building interaction via smart zoning of thermostatic loads: A switched self-tuning approach. *Appl. Energy* **2018**, *231*, 1246–1258. [CrossRef]
19. Oldewurtel, F.; Parisio, A.; Jones, C.N.; Gyalistras, D.; Gwerder, M.; Stauch, V.; Lehmann, B.; Morari, M. Use of model predictive control and weather forecasts for energy efficient building climate control. *Energy Build.* **2012**, *45*, 15–27. [CrossRef]
20. Dong, B.; Lam, K.P. A real-time model predictive control for building heating and cooling systems based on the occupancy behavior pattern detection and local weather forecasting. In *Building Simulation*; Springer: Berlin, Germany, 2014; Volume 7, pp. 89–106.
21. Pcolka, M.; Zacekova, E.; Robinett, R.; Celikovsky, S.; Sebek, M. Economical nonlinear model predictive control for building climate control. In Proceedings of the American Control Conference (ACC), Portland, OR, USA, 4–6 June 2014; pp. 418–423.
22. Klaučo, M. *Modeling of the Closed-Loop System with a Set of PID Controllers*; Slovak University of Technology: Bratislava, Slovakia, 2016.
23. Oldewurtel, F.; Sturzenegger, D.; Morari, M. Importance of occupancy information for building climate control. *Appl. Energy* **2013**, *101*, 521–532. [CrossRef]
24. Sturzenegger, D.; Gyalistras, D.; Morari, M.; Smith, R.S. Model Predictive Climate Control of a Swiss Office Building: Implementation, Results, and Cost–Benefit Analysis. *IEEE Trans. Control Syst. Technol.* **2016**, *24*, 1–12. [CrossRef]
25. Killian, M.; Kozek, M. Implementation of cooperative Fuzzy model predictive control for an energy-efficient office building. *Energy Build.* **2018**, *158*, 1404–1416. [CrossRef]
26. EnergyPlus Official Website. Available online: https://energyplus.net (accessed on 21 June 2017).
27. Baldi, S.; Michailidis, I.; Ravanis, C.; Kosmatopoulos, E.B. Model-based and model-free "plug-and-play" building energy efficient control. *Appl. Energy* **2015**, *154*, 829–841. [CrossRef]
28. Baldi, S.; Karagevrekis, A.; Michailidis, I.T.; Kosmatopoulos, E.B. Joint energy demand and thermal comfort optimization in photovoltaic-equipped interconnected microgrids. *Energy Convers. Manag.* **2015**, *101*, 352–363. [CrossRef]
29. Satyavada, H.; Baldi, S. An integrated control-oriented modelling for HVAC performance benchmarking. *J. Build. Eng.* **2016**, *6*, 262–273. [CrossRef]

30. Wu, S.; Sun, J.Q. A physics-based linear parametric model of room temperature in office buildings. *Build. Environ.* **2012**, *50*, 1–9. [CrossRef]

31. Baldi, S.; Yuan, S.; Endel, P.; Holub, O. Dual estimation: Constructing building energy models from data sampled at low rate. *Appl. Energy* **2016**, *169*, 81–92. [CrossRef]

32. Maasoumy, M.; Sangiovanni-Vincentelli, A. Total and peak energy consumption minimization of building HVAC systems using model predictive control. *IEEE Des. Test Comput.* **2012**, *29*, 26–35. [CrossRef]

33. Lauro, F.; Longobardi, L.; Panzieri, S. An adaptive distributed predictive control strategy for temperature regulation in a multizone office building. In Proceedings of the 2014 IEEE International Workshop on Intelligent Energy Systems (IWIES), San Diego, CA, USA, 5–8 October 2014; pp. 32–37.

34. Fanger, P.O. Calculation of thermal comfort, Introduction of a basic comfort equation. *ASHRAE Trans.* **1967**, *73*, III–4.

35. Michailidis, I.T.; Baldi, S.; Pichler, M.F.; Kosmatopoulos, E.B.; Santiago, J.R. Proactive control for solar energy exploitation: A german high-inertia building case study. *Appl. Energy* **2015**, *155*, 409–420. [CrossRef]

36. Rohles, J.; Frederick, H. Thermal sensations of sedentary man in moderate temperatures. *Hum. Fact.* **1971**, *13*, 553–560. [CrossRef] [PubMed]

37. Nguyen, T.A.; Aiello, M. Energy intelligent buildings based on user activity: A survey. *Energy Build.* **2013**, *56*, 244–257. [CrossRef]

energies

MDPI

Article

Prospects of a Meshed Electrical Distribution System Featuring Large-Scale Variable Renewable Power

Marco R. M. Cruz [1], **Desta Z. Fitiwi** [2], **Sérgio F. Santos** [1], **Sílvio J. P. S. Mariano** [3] and **João P. S. Catalão** [1,4,5,*]

1   C-MAST, University of Beira Interior, 6201-001 Covilhã, Portugal; marco.r.m.cruz@gmail.com (M.R.M.C.); sdfsantos@gmail.com (S.F.S.)
2   Energy and Environment Department, Economic and Social Research Institute, Dublin, Ireland; destinzed@gmail.com
3   Instituto de Telecomunicações and University of Beira Interior, 6201-001 Covilhã, Portugal; sm@ubi.pt
4   INESC TEC and the Faculty of Engineering, University of Porto, 4200-465 Porto, Portugal
5   INESC-ID, Instituto Superior Técnico, University of Lisbon, 1049-001 Lisbon, Portugal
*   Correspondence: catalao@fe.up.pt

Received: 27 October 2018; Accepted: 1 December 2018; Published: 4 December 2018

Abstract: Electrical distribution system operators (DSOs) are facing an increasing number of challenges, largely as a result of the growing integration of distributed energy resources (DERs), such as photovoltaic (PV) and wind power. Amid global climate change and other energy-related concerns, the transformation of electrical distribution systems (EDSs) will most likely go ahead by modernizing distribution grids so that more DERs can be accommodated. Therefore, new operational strategies that aim to increase the flexibility of EDSs must be thought of and developed. This action is indispensable so that EDSs can seamlessly accommodate large amounts of intermittent renewable power. One plausible strategy that is worth considering is operating distribution systems in a meshed topology. The aim of this work is, therefore, related to the prospects of gradually adopting such a strategy. The analysis includes the additional level of flexibility that can be provided by operating distribution grids in a meshed manner, and the utilization level of variable renewable power. The distribution operational problem is formulated as a mixed integer linear programming approach in a stochastic framework. Numerical results reveal the multi-faceted benefits of operating distribution grids in a meshed manner. Such an operation scheme adds considerable flexibility to the system and leads to a more efficient utilization of variable renewable energy source (RES)-based distributed generation.

Keywords: electrical distribution systems; meshed topology; mixed integer linear programing; stochastic programming; variable renewable power; flexibility option

1. Introduction

1.1. Framework and Motivation

Distribution power systems are experiencing massive transformations that are buoyed by the increasing need to integrate more variable renewable-type distributed generations (DGs). This means that distribution grids will be equipped with the necessary tools to enable bidirectional power flows, which is contrary to their traditional setup [1–7]. Also, such a massive transformation needs to be accompanied by new operational schemes. In other words, new operational strategies should be crafted and widely used in order to increase the degree of flexibility in the distribution systems, and hence the penetration of renewables, such as photovoltaic (PV) and wind power. This is due to the fact that the traditional radial network operation strategy may not be sufficient to accommodate the increasing penetration of renewables and for their efficient utilization.

In this context, smart grids are one of the most promising solutions that enable large-scale integration of variable renewable energy sources (vRESs) at a distribution level [8–12]. However, the scale of transformation that is required to "smartify" existing grids means that the whole process may be costly and, most importantly, slow. In other words, the smartification process will not happen overnight; it will, rather, involve a series of time-consuming and expensive upgrades to existing network infrastructures. Hence, the impact of smart grids would only be felt in the long-run when they have fully materialized.

Most of the traditional distribution networks are meshed by design, but they are operated radially only due to technical limitations that are mainly related to system protection. These limitations are discussed in [13]. This means that some tie-lines (also known as switches) are kept open so that the grid's topology remains radial. Thus far, it has been easier to operate and protect a radial topology [13–16].

The lines that are normally open in radial network systems are only used in emergency situations, e.g., situations of fault or power supply failure. The main purpose of this is to enhance the reliability of power delivery, i.e., some of the tie-lines that are normally open are closed to re-route power flows so that the amount of load that is shed is minimized. However, the radiality of the network system is maintained at all times, regardless of the operational situations that happen in the system. The good news is that, in well-planned distribution networks, contingencies or emergency situations are rare phenomena.

As previously mentioned, one strategy worth considering is the operation of distribution networks in a meshed topology. This type of topology is contrary to what is well-established; that is, radially operating distribution systems. However, with the technological advances that are now available, and expected to happen in the near future in an even more accelerated manner, it is possible to deal with all of the inherent limitations of the meshed operation of distribution networks. Given its multi-faceted benefits, the so-called meshed topology is expected to be a normal operation scheme for distribution grids in the future. However, this does not mean that a radial topology would be completely abolished; there may be cases where this would make more sense from an economic and a technical standpoint.

The advantages that meshed distribution systems have are the reduction of power losses, improved voltage profiles, more flexibility, the capability to deal with high electricity demand growth, and the enhancement of power quality (PQ) [17]. Furthermore, in meshed distribution systems with no integrated DGs, the distribution of power flows among parallel paths can potentially decrease the stress on the entire network system, and possibly defer grid-related investments. This can be achieved only by optimizing the loops in tie-lines in the distribution system. When DGs are appropriately allocated in such systems, they can bring about several benefits, such as a reduction of power losses, a better voltage profile, and also an investment deferral as a result of reduced congestion in the network components (feeders and transformers) [17].

Likewise, a meshed topology can provide similar benefits to those of DGs. The combination of both can potentially enhance distribution system reliability and the quality of the power that is delivered to end-users. The negative aspects that are associated with DG integration are the possible increase in short circuit currents and, hence, a possible need for modification of protection devices' settings [17].

Because of this, international standards determine the immediate disconnection of DGs from the distribution system in case of faults so that conventional protection devices can act properly. Similarly, a meshed topology also shares this issue. However, technological advances make it easier to switch from meshed to a radial topology in case of fault or vice versa, allowing us to reap the benefits of a meshed topology. For example, locally generated renewable power can be efficiently utilized under a meshed topology, which would otherwise have been curtailed in the traditional network setup.

There are a set of technologies that could be used to exploit a meshed network topology, and minimize the limitations of such a topology. For instance, when a fault occurs in the system, fast de-loopers can be deployed to quickly switch from a meshed to a radial topology so that conventional protection devices can properly act. Another enabling technology with regards to a meshed topology's operation is Fault Current Limits (FCLs) [18]. Generally, the operation of distribution networks in a meshed manner may become the norm in the near future.

1.2. Literature Review

The large-scale integration of vRESs has brought about a set of challenges that require attention and action. The main challenges involve protection schemes, voltage regulation as a result of fluctuations induced by vRESs, voltage sags and/or rises, and network congestion [19]. Such issues are exacerbated by the increasing integration of vRESs in distribution networks because these are designed for unidirectional power flows. However, operating distribution grids in a meshed manner can alleviate some of these issues, and bring about a number of benefits; for example, in terms accommodating more vRES power, which is an important aspect given the growing global concerns about climate change. However, the prospects of a meshed operational scheme have not been adequately explored in the literature.

A comparative study between a meshed operation and the reinforcement of distribution networks was conducted in [20]. The study involves the comparison of results from enumeration, constraints, and loops analysis methods. The authors in [21] develop a model that estimates the maximum penetration of DGs based on a steady state analysis. The approach uses some elements of an optimal power flow analysis and bus voltage and current flow limits to estimate the maximum allowable DG penetration at each node of the considered system. The authors conclude that a meshed topology may be a good alternative to host large-scale DG power. Furthermore, the authors in [22] propose a methodology for allocating conventional DGs in a distribution system, and evaluate their impacts on the distribution system. For the analysis, they have considered a meshed operational scheme and a voltage sensitivity index to quantify the operability of the system. However, their analysis is based on the integration of conventional sources of energy into distribution systems. In [23], the authors provide an extensive analysis of optimum power flows when operating distribution networks in a radial and a meshed manner. The analysis is carried out considering reactive power compensation devices and DGs. In [24], the authors develop an operational model for analyzing the prospects of a meshed distribution network topology, which is based on circuits composed of a resistor, inductor, and capacitor (RLC). Reference [25] provides a steady state analysis of a meshed distribution system featuring DGs, and is based on iterative load-flow calculations. However, the analysis of the existing literature, reviewed here, is based on conventional DGs under a meshed operational scheme. To the best of our knowledge, the topic of integrating vRESs in tandem with meshed operation of distribution systems has not been addressed in the literature. Hence, this is the main focus of the current work. The argument provided here is that electrical distribution systems can be operated in a meshed topology under normal situations. Additionally, they can be equipped with advanced, and even currently available, technologies that temporarily enable a smooth automated transition to a radial topology in case of a contingency, and back to the preferred topology when the fault is cleared. This way, one can take full advantage of the meshed operation of the distribution network, which eventually leads to reduced losses, improved voltage profiles, and a more efficient management of locally produced vRES power. The meshed operational scheme can also have benefits in terms of network-related investments. The more distributed nature of power flows in the meshed topology would mean lower stress (congestion) in the whole system, reducing the need for network upgrades. Note that existing switches and loops in distribution systems can be effectively used to develop an optimal meshed topology.

1.3. Contributions and the Paper's Organization

The increasing penetration of vRESs into distribution networks unfortunately leads to some technical challenges. Appropriate tools and methods need to be developed and deployed to address these challenges. One example is exploring different operating strategies, which forms the core analysis of the subject matter that is contained in our current work. The main enabling mechanisms of a meshed operational scheme in a distribution network are discussed.

An appropriate mathematical formulation is developed to support our analysis. The formulation is an operational model that is based on a least-cost optimization formulated under a stochastic programming framework to account for the uncertainty that arises from various sources, such as renewable power production and electricity demand. The analysis encompasses an optimal network topology, losses, voltage profiles, costs, and other operational variables of the considered distribution network. The resulting model is of a stochastic mixed integer linear programming type, whose objective function is minimized to achieve the optimal operation of a system that features large quantities of vRES power, while respecting a number of technical, economic, and environmental constraints. An extensive analysis is performed with respect to the flexibility that a meshed topology can provide, and the impact on the integration and utilization levels of vRESs.

This work is organized as follows. The mathematical model used in this work is described in Section 2. Section 3 presents details of the case studies that are considered in this work, and discusses the numerical results that were obtained. Section 4 provides concluding remarks and further insights.

2. Mathematical Model

In this section, we provide a broad description of the developed stochastic optimization model, which is of a mixed integer linear programming nature. The model is used to conduct an analysis of the optimal operation of meshed distribution systems with large-scale renewable integration. The meshed operation is analyzed in terms of the use level of locally produced vRES power without adversely affecting the operation of the system. A linearized alternating current (AC) power flow model is used. This model guarantees a correct balance between accuracy and computational requirements. Note that the acronyms used in the mathematical formulation are presented in Appendix A.

2.1. Objective Function

The main objective is to minimize the total costs of operating the considered distribution system. These costs are associated with the operating costs in the system, namely the cost of energy not supplied, the costs of emissions, and the cost of power generation using conventional and renewable power sources.

$$Minimize\ TOC = TEC + TENSC + TEmiC \tag{1}$$

Equation (1) minimizes the *TOC*, which represents the total expected cost in the system.

The first term in (1) represents the expected costs of producing energy using renewable technologies (solar and wind in this case), and purchasing energy from the upstream network as in (2). The two terms in (2) are calculated by (3) and (4), respectively.

$$TEC = EC^{vRES} + EC^{SS} \tag{2}$$

$$EC^{vRES} = \sum_{s\in\Omega^s} \rho_s \sum_{h\in\Omega^h} \sum_{g\in\Omega^g} OC_g P^{vRES}_{g,i,s,h}, \tag{3}$$

$$EC^{SS} = \sum_{s\in\Omega^s} \rho_s \sum_{h\in\Omega^h} \sum_{\varsigma\in\Omega^\varsigma} \lambda^\varsigma_h P^{SS}_{\varsigma,s,h}. \tag{4}$$

Regarding the second term of (1), *TENSC* represents the cost of energy not supplied. This term is based on the calculation of the active and reactive power that was not supplied, and is given by Equation (5).

$$TENSC = \sum_{s\in\Omega^s} \rho_s \sum_{h\in\Omega^h} \sum_{i\in\Omega^i} \left(v^P_{s,h} P^{NS}_{i,s,h} + v^Q_{s,h} Q^{NS}_{i,s,h} \right) \tag{5}$$

The terms $v_{s,h}^P$ and $v_{s,h}^Q$ are defined as penalty factors. They correspond to penalty terms that are associated with any active and reactive power that is shed. These must be set to sufficiently high values to avoid unnecessary load shedding. Finally, the term $TEmiC^{vRES}$ represents the expected cost of emissions. These emissions are related to energy production from renewable sources, as well as conventional ones, and that of energy purchased from the upstream network. This term is defined by:

$$TEmiC = TEmiC^{vRES} + EmiC^{SS}. \tag{6}$$

The corresponding terms in (5) are expressed by:

$$TEmiC^{vRES} = \sum_{s\in\Omega^s} \rho_s \sum_{h\in\Omega^h} \sum_{g\in\Omega^g} \sum_{i\in\Omega^i} \lambda^{CO_2} ER_g^{vRES} P_{g,i,s,h}^{vRES}, \tag{7}$$

$$EmiC^{SS} = \sum_{s\in\Omega^s} \rho_s \sum_{h\in\Omega^h} \sum_{\varsigma\in\Omega^\varsigma} \sum_{i\in\Omega^i} \lambda^{CO_2} ER_\varsigma^{SS} P_{\varsigma,s,h}^{SS}. \tag{8}$$

2.2. Constraints

Kirchhoff's current law must be enforced for active (9) and reactive (10) power flows. These ensure that the sum of incoming flows must be equal to the outgoing ones. These conditions must be respected at all times for safe operation of the system.

$$\sum_{g\in\Omega^g} P_{g,i,s,h}^{vRES} + P_{\varsigma,s,h}^{SS} + P_{i,s,h}^{NS} + \sum_{in,k\in\Omega^k} P_{k,s,h} - \sum_{out,k\in\Omega^k} P_{k,s,h} = PD_{s,h}^i + \\ \sum_{in,k\in\Omega^k} \tfrac{1}{2} PL_{k,s,h} + \sum_{out,k\in\Omega^k} \tfrac{1}{2} PL_{k,s,h}; \forall\varsigma\in\Omega^\varsigma; \forall\varsigma ei; kei, \tag{9}$$

$$\sum_{g\in\Omega^g} Q_{g,i,s,h}^{vRES} + Q_{\varsigma,s,h}^{SS} + Q_{i,s,h}^{NS} + \sum_{in,k\in\Omega^k} Q_{k,s,h} - \sum_{out,k\in\Omega^k} Q_{k,s,h} = QD_{s,h}^i + \\ \sum_{in,k\in\Omega^k} \tfrac{1}{2} QL_{k,s,h} + \sum_{out,k\in\Omega^k} \tfrac{1}{2} QL_{k,s,h}; \forall\varsigma\in\Omega^\varsigma; \forall\varsigma ei; kei. \tag{10}$$

On the left-hand side of Equation (9), we can see that the active power flows from the renewable power generation as well as the power that is injected at the substation. On the other side of the equation, we have the power flow that is associated with the demand and the losses (treated here as fictitious loads). The same principles apply to the reactive power flow shown in (10).

Kirchhoff's voltage law must also be considered. This restriction governs the power flow in the feeders, which are represented by linearized power flow equations considering two practical assumptions. The first one states that the voltage magnitude is essentially close to the nominal value V_{nom}. The second one is related to the difference of voltage angles θ_k. For security systems, this difference has to be as small as possible, which leads to a trigonometric approximation $\sin\theta_k \approx \theta_k$ and $\cos\theta_k \approx 1$. Considering these two simplifying assumptions, the active and reactive AC power flow equations can be linearized, and represented as in:

$$\left| P_{k,s,h} - \left(V_{nom}\left(\Delta V_{i,s,h} - \Delta V_{j,s,h} \right) g_k - V_{nom}^2 b_k \theta_{k,s,h} \right) \right| \leq MP_k(1 - u_{k,h}), \tag{11}$$

$$\left| Q_{k,s,h} - \left(-V_{nom}\left(\Delta V_{i,s,h} - \Delta V_{j,s,h} \right) b_k - V_{nom}^2 g_k \theta_{k,s,h} \right) \right| \leq MQ_k(1 - u_{k,h}), \tag{12}$$

$$\text{where } \Delta V^{min} \leq \Delta V_{i,s,h} \leq \Delta V^{max}. \tag{13}$$

In relation to the power flows in each branch, these cannot exceed the maximum transfer capacity:

$$P_{k,s,h}^2 + Q_{k,s,h}^2 \leq u_{k,h}(S_k^{max})^2. \tag{14}$$

The active and reactive power losses in each branch are algebraically represented by:

$$PL_{k,s,h} = R_k \left(P_{k,s,h}^2 + Q_{k,s,h}^2 \right) / V_{nom}^2, \tag{15}$$

$$QL_{k,s,h} = X_k \left(P_{k,s,h}^2 + Q_{k,s,h}^2 \right) / V_{nom}^2. \tag{16}$$

Note that Equations (14)–(16) are easily linearized using a piecewise linearization approach, which is common in the literature. Further explanation of the piecewise linearization can be found in Appendix B.

The active and reactive power limits of conventional as well as vRESs are also considered as constraints. Such constraints related to vRESs are given by (17) and (18):

$$P_{g,i,s,h}^{vRES,min} \leq P_{g,i,s,h}^{vRES} \leq P_{g,i,s,h}^{vRES,max}, \tag{17}$$

$$-\tan\left(\cos^{-1}(pf_g)\right) P_{g,i,s,h}^{vRES} \leq Q_{g,i,s,h}^{vRES} \leq \tan\left(\cos^{-1}(pf_g)\right) P_{g,i,s,h}^{vRES}. \tag{18}$$

where pf_g is the power factor of generator g.

The reactive power injected or withdrawn at a substation in the system is subject to a minimum and a maximum level as in (18). This is motivated by security concerns.

$$-\tan\left(\cos^{-1}(pf_{ss})\right) P_{\varsigma,s,h}^{SS} \leq Q_{\varsigma,s,h}^{SS} \leq \tan\left(\cos^{-1}(pf_{ss})\right) P_{\varsigma,s,h}^{SS}, \tag{19}$$

Note that the voltage angle difference $\theta_{k,s,h}$ is defined as $\theta_{k,s,h} = \theta_{i,s,h} - \theta_{j,s,h}$. In this case, i and j belong to the same branch k.

3. Results

3.1. Data and Assumptions

In this work, we use a standard 119-bus distribution system to perform the required analysis. The schematic diagram of this system is shown in Figure 1. The main data of the considered system are summarized in Table 1. Further information about the test system and data can be found in [3]. The size and location of vRESs are adapted from [3], as can be seen in Table 2. More of the data-related assumptions that were made in this analysis are presented in Tables 3–5. Further assumptions are summarized as follows:

- The operational analysis is based on a 24-h period, subdivided on an hourly basis.
- The maximum voltage deviation at each bus is set to ±5% of the nominal value (which, in this case, is 11 kV).
- In all simulations, the substation is treated as the reference node, in which both the voltage deviation and the angle are set to zero.
- The number of partitions considered for linearizing quadratic terms is 5, which is in line with the findings in [26].

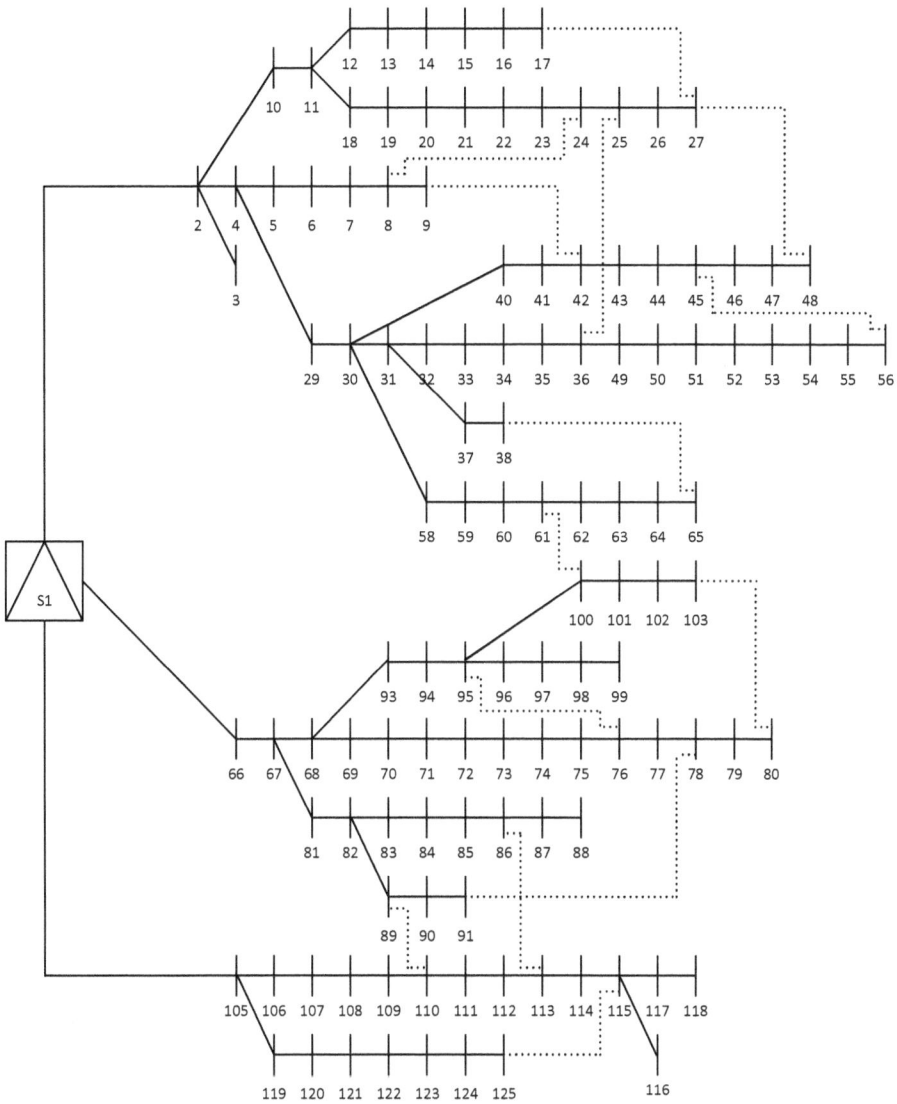

Figure 1. A schematic diagram of the 119-bus test system.

Table 1. General system data.

Parameter Description	Parameter Setting
Nominal voltage	11 kV
Active power demand	22,709.720 kW
Reactive power demand	17,041.068 kVAr
Base case system losses	1298.090 kW
Minimum voltage of the base case system (which occurs at bus 116)	0.8783 p.u.

Table 2. The size and location of wind- and photovoltaic (PV)-type distributed generations (DGs).

Bus	Wind (MW)	PV (MW)
14	1	0
21	1	0
24	1	0
25	0	1
29	0	1
32	1	0
33	1	0
35	0	1
37	1	0
38	1	0
42	1	0
43	0	1
44	1	1
52	1	1
53	1	0
56	1	0
61	1	0
69	1	0
73	1	1
74	1	0
77	1	1
79	0	1
82	1	0
83	1	0
84	0	1
85	1	0
89	1	0
96	1	0
100	1	1
101	0	1
106	0	1
108	1	0
112	1	1
114	1	1
116	1	1
117	0	1
119	0	1
121	1	0

Table 3. Further data-related assumptions.

Parameter	Setting
pf^{ss}	0.8
pf^{vRES}	0.95
ER^{ss}	0.4 tCO$_2$e/MWh
λ^{CO_2}	15 /tCO$_2$e
$v_{s,h}^{P}$	3000 /MW
$v_{s,h}^{Q}$	3000 /MVAr
MP_k, MQ_k	20

Table 4. The cost of electricity generation from renewable sources and emission rates.

DG Type	Variable Cost (€/MWh)	Emission Rates of DGs (tCO$_2$e/MWh)
Solar	40	0.0584
Wind	20	0.0276

Table 5. The maximum transfer capacity in feeders.

Feeders	Maximum Transfer Capacity (A)
{(1, 2); (2, 4); (1, 66); (66, 67)}	1200
{(4, 5); (5, 6); (6, 7); (4, 29); (29, 30); (30, 31); (67, 68); (67, 81); (81, 82); (1, 105); (105, 106); (106, 107)}	800
Remaining feeders	400

Our work involves power productions using variable renewable sources, such as wind and solar. The power outputs from these resources are subject to high-level uncertainty and variability. Demand is also variable (say throughout the course of the day), even if it can be fairly predicted more accurately than a variable renewable power output. The stochastic nature of our work arises as a result of these issues. Therefore, we have handled such stochastic parameters via a stochastic programming framework that accounts for the most plausible states of these parameters at a given future time, each of which is associated with a probability. Over the considered operational period (which in our case is 24-h long), such states collectively form scenarios, which are jointly considered in the optimization process.

In other words, the stochastic nature of RES power outputs and demand is accounted for by considering an adequate number of scenarios for each. Therefore, the power production profiles of wind- and solar-type DGs, as well as the demand profile, are assumed to be uniform throughout the system. The uncertainty associated with solar and wind power generations are taken into account by considering three different scenarios for each uncertain parameter. Demand uncertainty is also taken into account by considering six different scenarios each for residential- and industrial-type consumers. It should be noted that each scenario represents an hourly profile. The combination of these individual scenarios (which, in this case, is 81) results in the creation of the final set of scenarios that is used in our studies.

3.2. Numerical Results

As stated above, the analysis is carried out to study the operational flexibility that can be provided by operating vRES-rich distribution grids in a meshed manner. In addition, the analysis includes the impact of such a scheme on the use and integration of vRESs.

A total of six case studies are considered, designated as Case A to Case F. Case A is the Base Case, which neither considers network reconfiguration nor meshed operation. In Case B, network reconfiguration is allowed but always while maintaining a radial topology. Cases C to F all consider a meshed operational scheme, but with different levels of meshing (30% in Case C, 60% in case D, 80% in Case E, and 100% in Case F). The network configurations for the last four cases are presented in Figure 2. Note that meshing the distribution network makes use of existing tie-lines. For Cases B to F, the upper and lower voltage boundaries have been enforced. Table 6 shows the total expected cost, along with a breakdown of this cost and the total expected power losses in the system.

Table 6. The total expected costs of the objective function and power losses. PNS: Power Not Served.

	Case A	Case B	Case C	Case D	Case E	Case F
Total Cost (€)	32,217.38	27,215.55	24,634.12	18,458.99	16,937.63	15,664.99
Energy Cost (€)	30,349.82	26,629.07	24,103.04	17,979.25	16,501.48	15,265.23
Emission Cost (€)	1219.56	557.47	513.63	472.96	436.15	399.76
PNS Cost (€)	647.99	29.01	17.45	6.78	0.00	0.00
Power Loss (MW)	20.25	9.39	8.01	7.21	6.47	5.73
Power Loss (MVar)	14.11	6.13	4.67	3.97	3.24	2.49

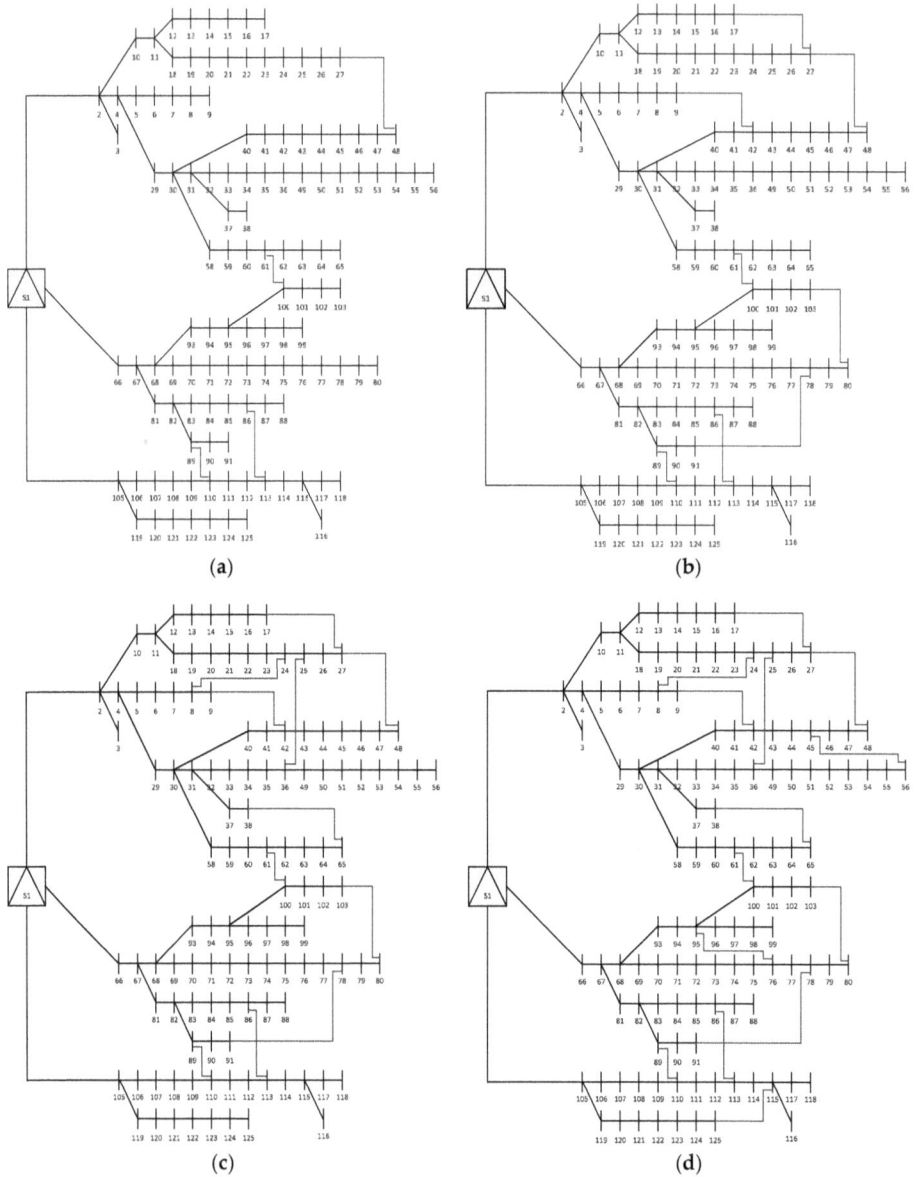

Figure 2. A schematic diagram of the meshed systems associated with Case C to Case F. (**a**) The 30% meshed network topology for Case C; (**b**) the 60% meshed network topology for Case D; (**c**) the 80% meshed network topology for Case E; and (**d**) the 100% meshed network topology for Case F.

Among the considered cases, the Base Case has the highest value of total costs, as expected. This is because all of the energy required in the system is imported through the substation. The energy mix associated with the Base Case can be seen in Figure 3a. Apart from the costs, the power losses in the system are also the highest among those computed in the remaining cases.

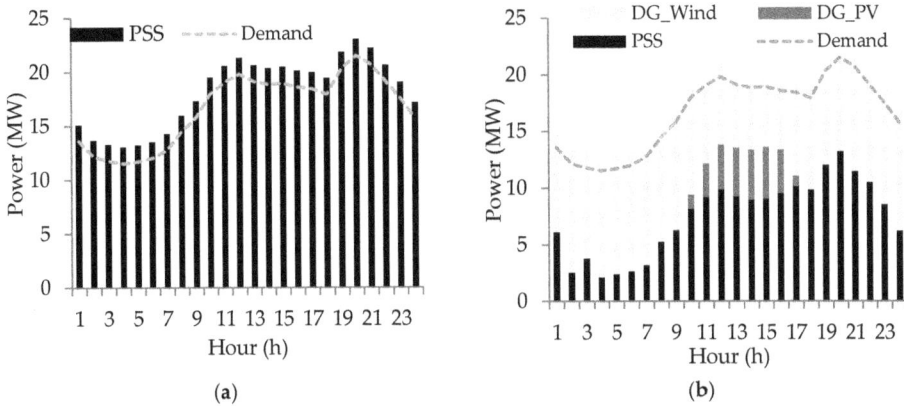

Figure 3. Energy mixes in the (**a**) Base Case and (**b**) Case B. PSS: Power from Substation; DG: distributed generation.

In Case B, the expected energy costs are reduced by 12%, the expected emissions costs by 54%, and the expected Power Not Served (PNS) costs by 96%. Overall, this translates into a reduction of 16% in expected total cost in the system. The decreases registered in the expected energy and emission costs are mainly due to the locally produced vRES power that is cheaper and "cleaner". The active and reactive power losses in the system are reduced on average by 54% and 70%, respectively. This is as a result of the combined effect of the DG integration and the optimal network reconfiguration. Most of the demand is met by locally generated power, which does not require heavy utilizations of existing feeders, and hence results in reduced losses. It has been proven that an optimal reconfiguration also reduces losses in the system. Figure 3b shows Case B's energy mix. In this figure, it can be seen that this case has 60.4% of the demand met by vRES power (of which 6.6% comes from solar-type and 53.9% from wind-type DGs).

Cases C through F are the ones that represent a system that is operated in a meshed network topology, but with increasing levels of "meshedness". In Case C, a 24% greater reduction in the expected total cost is observed, as a result of reductions in the individual cost terms (energy, emission, and PNS costs), in comparison with that of the Base Case. Active and reactive power losses also see reductions on average by 60% and 77%, respectively. With the increase in the "meshedness" level of the network, the reductions become more pronounced. In Case C, the percentage of demand covered by vRES power is 69.7% (of which 7.3% comes from solar and 62.4% from wind). In comparison to the radial topology in the Base Case, even the less-meshed topology sees further improvement in the utilization level of vRES power production. A further observation is the fact that even a weakly meshed distribution network (with a 30% connectedness index) shows an improvement of 9.3% in terms of vRES power generation compared to that of an optimally reconfigured radial topology.

In Cases D and E, where the "meshedness" levels are 60% and 80%, respectively, one can observe 43% and 47% overall cost reductions, respectively. In comparison to that of the Base Case, these can be regarded as significant improvements, and these generally show the favorable impact of meshed system operation. In Case E, the PNS costs are reduced by 100%. This can be explained by the fact that meshing the grid routes vRES power to where it is consumed. This would otherwise be shed in the radial (or weakly meshed) topology. As a result, the share of renewables in the total consumption in the above two cases (i.e., Cases D and E) amounts to 73.6% and 75.1%, respectively.

The last case—Case F—(where all branches are connected, creating a completely meshed network) yields the best operational results among the considered cases. Compared to the Base Case, a 51% reduction in overall cost can be seen. In terms of individual cost terms, the reductions are 50% in expected energy costs, 67% in emission costs, and 100% in expected PNS costs. System-wide average losses are slashed down by 72% (active) and 88% (reactive). Regarding the energy mix, the fully meshed network, i.e., Case F, has a total of 75.8% of the total energy demand met by vRES energy (out of which 10.7% comes from solar-type and 65.1% from wind-type DGs). From Case A to Case F, one can easily notice the reductions in terms of energy imported from upstream (see in Table 6). In Case F, the entire system operates in near island mode (see hours 4 and 5, in which only 3% of demand in these hours is covered by importing power from the upstream network). Also, numerical results highlight that a fully meshed topology increases the utilization level of vRESs power generation by 15.4% compared to that of an optimally reconfigured radial topology. This translates into an approximately 42% decrease in the overall system cost, and 44% and 99% reductions in terms of expected energy and emission costs, respectively, as compared to that of a reconfigured radial topology, which is significant. The share of renewable power in the final energy consumption is as high as 75.8% in the case that incorporates a strongly meshed network, which is again noteworthy.

Figure 4a–d shows the energy mixes corresponding to the meshed cases, from low to a more complex meshed topology, respectively. The results in these figures reveal interesting variations in the utilization levels of vRES power productions during the 24-h period. It is also possible to see a decrease in the energy purchased from the upstream network (PSS) throughout the various case studies.

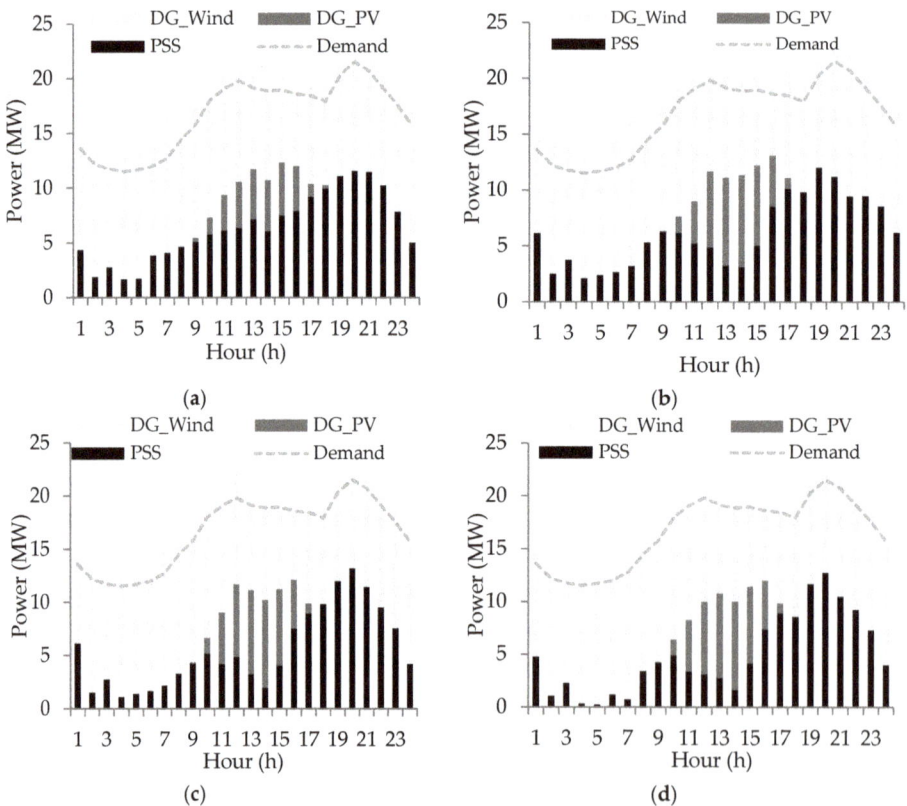

Figure 4. Energy mixes for different case studies. (**a**) Case C energy mix (30% meshed network); (**b**) Case D energy mix (60% meshed network); (**c**) Case E energy mix (80% meshed network); and (**d**) Case F energy mix (100% meshed network).

With regard to energy losses, the average profile of active power losses during the considered 24-h period of each case is shown in Figure 5. The results are in accordance with Table 6, dropping from Case B to Case F, as has already been mentioned before. In Cases C to F, losses decrease within an interval, since, in addition to the DGs being near the loads, there are also now in some sections of the network smaller paths to be "traveled", resulting in a losses decrease.

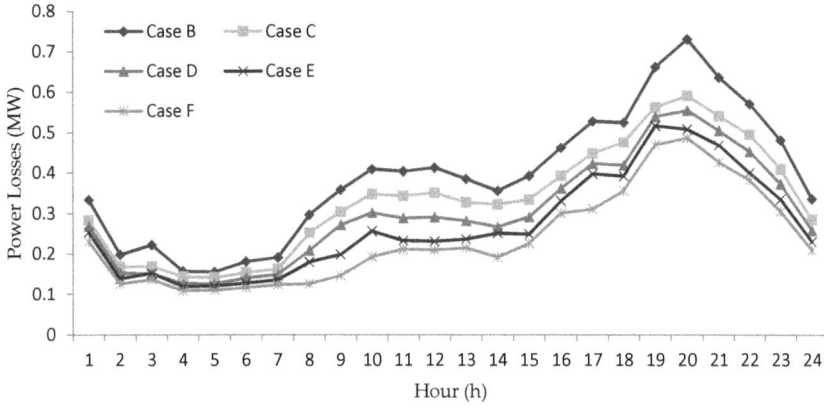

Figure 5. The power losses in the network associated with each case.

Figure 6 shows the average voltage profile corresponding to each case. The voltage deviation profile in Case A is the only one in which the deviations in some nodes surpasses the lower bound. In the remaining cases, where DGs are already integrated, all voltage deviations are significantly improved, and largely remain within the permissible range. In Case B as well as in the cases that involve network meshing (i.e., Cases C through F), voltage deviations do not show significant differences. In the figure, detailed voltage deviations for the nodes from 41 to 53 can be seen in the section that is zoomed out. In this particular section, we can see minor improvements in the voltage deviation, especially from Cases C through F. Generally, the case that has the most meshed network (i.e., Case F) has the best voltage deviation portfolio among the cases.

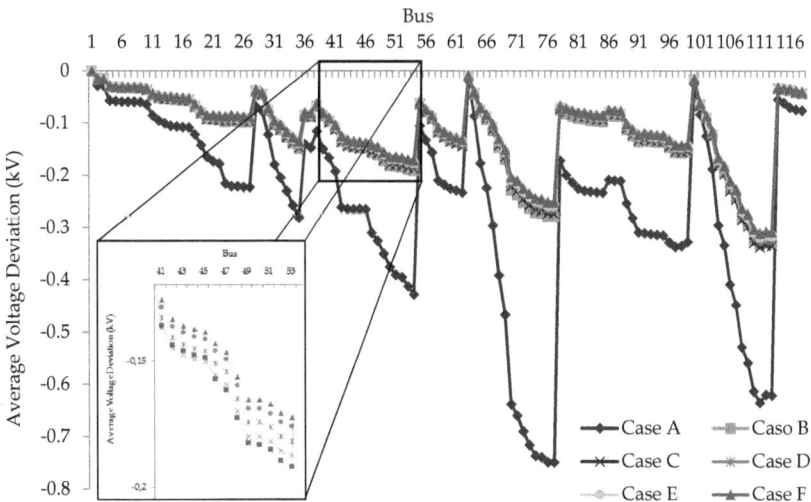

Figure 6. The average voltage deviation in all cases.

Total solar and wind power productions by node are shown in Figure 7a,b, respectively. In these figures, it is possible to observe the increased vRES power generation as one moves from Case B to Case F at each node. From the results in these figures, along with those in Figure 2, we can see the complementarity of meshed operation and renewable integration, in which a higher network meshing leads to a higher network flexibility and, hence, a greater increase in renewable integration.

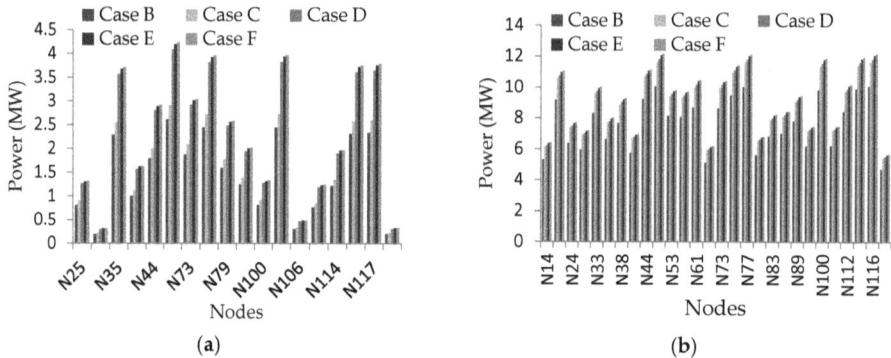

(a)

(b)

Figure 7. Variable renewable energy sources (vRES) outputs by node. (**a**) Solar power outputs by node (**b**) Wind power outputs by node.

Largely, the results obtained in the case studies, but especially in Case F, point out the immense contributions of the meshed operational scheme in terms of increasing system flexibility and the efficient utilization of vRESs in the system.

4. Conclusions

The work in this paper has explored the prospects of operating distribution network systems in a meshed topology, as opposed to the conventionally known radial operation. Furthermore, the contributions of a meshed network topology are studied in terms of enhancing system flexibility and its potential to increase the integration and efficient utilization of vRES power generation. To accomplish this, a stochastic mixed integer linear programming (SMILP) optimization model has been developed with a reasonably large scale distribution network as a test system. A linearized AC power flow is used, and the operational problem is formulated as a least-cost optimization while satisfying a number of technical, economic, and environmental constraints. Numerical results from the cases considered show that adopting a meshed network topology as a mainstream operational strategy for distribution systems has considerable benefits. Generally, a more meshed network leads to a better utilization of locally produced vRES power, and, hence, a higher share of renewable power. Even a weakly meshed distribution network (i.e., the case with a 30% connectedness index) results in a 9.3% increase in vRES power absorbed by the system during the considered time period. For the fully meshed topology case, the increase in the utilization level of vRES power amounts to 15.4% compared to that of an optimally reconfigured radial topology. This translates into a 42% decrease in the overall system cost. Also, the share of renewable power in the final energy consumption is as high as 75.8% in the case that incorporates a strongly meshed network, which is again noteworthy. Most importantly, all these improvements come without creating any undesirable effect on the operation of the considered distribution system. Instead, the average voltage profile is further enhanced, and the average power losses are significantly lowered. The results generally reveal the multi-faceted contributions and viability of a meshed operational strategy. It has been verified that this strategy adds valuable flexibility to distribution systems that are rich in vRES-based distribution generations. Such added system flexibility is an important asset to have for ensuring a more efficient utilization of variable renewable power generation in the system.

Energies **2018**, *11*, 3399

The authors will explore further opportunities for meshed operational schemes in the near future. In parallel to this, they intend to concentrate their focus on a detailed analysis in terms of its challenges when it comes to its practical implementation and propose possible solutions.

Author Contributions: Conceptualization, M.R.M.C., D.Z.F. and S.F.S.; Methodology, M.R.M.C., D.Z.F. and S.F.S.; Validation, D.Z.F. and J.P.S.C.; Writing, M.R.M.C., D.Z.F. and S.F.S.; Supervision, D.Z.F., S.J.P.S.M. and J.P.S.C.

Funding: João P. S. Catalão acknowledges the support by FEDER funds through COMPETE 2020 and by Portuguese funds through FCT, under Projects SAICT-PAC/0004/2015–POCI-01-0145-FEDER-016434, POCI-01-0145-FEDER-006961, UID/EEA/50014/2013, UID/CEC/50021/2013, UID/EMS/00151/2013, and 02/SAICT/2017–POCI-01-0145-FEDER-029803. Also, this publication has emanated from research supported in part by a research grant from Science Foundation Ireland (SFI) under the SFI Strategic Partnership Programme Grant number SFI/15/SPP/E3125. The opinions, findings and conclusions or recommendations expressed in this material are those of the authors and do not necessarily reflect the views of the Science Foundation Ireland.

Conflicts of Interest: The authors declare no conflict of interest.

Appendix A

The nomenclature that corresponds to the mathematical formulation that is presented in Section 2 can be seen in Table A1.

Table A1. Nomenclature.

Sets/Indices	Definitions	Sets/Indices	Definitions
i/Ω^i	Index/set of buses	$hh,h/\Omega^h$	Index/set of hourly snapshots
$g/\Omega^g/\Omega^{DG}$	Index/set of generators/vRES	s/Ω^s	Index/set of scenarios
k/Ω^k	Index/set of branches	$\varsigma/\Omega^\varsigma$	Index/set of substations
Parameters	**Definitions**	**Parameters**	**Definitions**
ER_g, ER^{SS}_ς	Emission rates of vRES and energy purchased, respectively (tCO$_2$e/MWh)	N_i, N_ς	Number of buses and substations, respectively
g_k, b_k, S^{max}_k	Conductance, susceptance, and flow limit of branch k (Ω, Ω, **MVA**)	V_{nom}	Nominal voltage (kV)
MP_k, MQ_k	Big-M parameters associated with active and reactive power flows through branch k	Z_{ij}, R_k, X_k	Impedances of branch i-j (Ω)
$OC_{g,i,s,h}$	Operation cost of unit energy production by vRES (€/MWh)	$\lambda^{CO_2e}_{s,h}$	Price of emissions (€/tons of CO$_2$ equivalent)
MP_k, MQ_k	Big-M parameters associated with active and reactive power flows through branch k	$\lambda^\varsigma_{s,h}$	Price of electricity purchased upstream (€/MWh)
$OC_{g,i,s,h}$	Operation cost of unit energy production by vRES (€/MWh)	ρ_s, π_w	Probability of hourly scenario s and weight (in hours) of hourly snapshot group h
$v^P_{s,h}, v^Q_{s,h}$	Penalty for active and reactive unserved power, respectively (€/MW, €/MVAr)	pf_{ss}, pf_g	Power factor of substation and DGs, respectively
L	Total number of linear segments	α_l, β_l	Slopes of linear segments
Variables	**Definitions**	**Variables**	**Definitions**
$PD^i_{s,h}, QD^i_{s,h}$	Active and reactive power demand at node i (MW, MVAr)	$P^{NS}_{i,s,h}$	Unserved power at node i (MW)
$P_{g,i,s,h}, Q_{g,i,s,h}$	Active and reactive power produced by vRES (MW)	$Q^{NS}_{i,s,h}$	Unserved power at node i (MW)
$P^{SS}_{\varsigma,s,h}, Q^{SS}_{\varsigma,s,h}$	Active and reactive power imported from the grid (MW)	V_i, V_j	Voltage magnitudes at nodes i and j (kV)
P_k, Q_k, θ_k	Active and reactive power flows, and voltage angle difference of link k (MW, MVAr, radians)	$u_{k,h}$	Utilization variables of existing lines
PL_k, QL_k	Active and reactive power losses (MW, MVAr)	θ_i, θ_j	Voltage angles at node i and j (radians)
$PL_{\varsigma,s,h}, QL_{\varsigma,s,h}$	Active and reactive power losses at substation ς (MW, MVAr)	$\lambda_{s,h'}$	Real-time price of electricity (€/MWh)
$p_{k,s,h,l}, q_{k,s,h,l}$	Step variables used in the linearization of quadratic flows (MW, MVAr)		
Functions	**Definitions**	**Functions**	**Definitions**
EC^{SS}_h	Expected cost of energy imported through the substation level	EC^{vRES}_h	Expected cost of energy produced by vRES
$EmiC^{vRES}_h$	Expected emission cost due to vRES power production (€)	$EmiC^{SS}_h$	Expected emission cost of energy imported through the substations (€)
TEC	Total expected costs of supplied energy (€)	$TENSC$	Total expected costs of energy not supplied (€)
$TEmiC$	Total expected costs of emissions (€)		

Appendix B

There are a number of ways to linearize the quadratic functions in (14) through (16), such as incremental, multiple choice, convex combination, and other approaches in the literature. However, as mentioned earlier, the quadratic terms are linearized via a piecewise linearization method, which is thoroughly described in [26,27]. For the sake of brevity, here, we only show the piecewise representations of $P^2_{k,s,h}$ and $Q^2_{k,s,h}$. Others follow the same procedure and involve similar sets of constraints.

Energies **2018**, *11*, 3399

The approach (which is based on a first-order approximation of the nonlinear curve) uses a sufficiently large number of linear segments, L. To this end, two non-negative auxiliary variables are introduced for each of the flows P_k and Q_k such that $P_k = P_k^+ - P_k^-$ and $Q_k = Q_k^+ - Q_k^-$. Note that these auxiliary variables (i.e., P_k^+, P_k^-, Q_k^+, Q_k^-) represent the positive and negative flows of P_k and Q_k, respectively. This helps one to consider only the positive quadrant of the nonlinear curve, resulting in a significant reduction in the mathematical complexity, and by implication the computational burden. In this case, the associated linear constraints are:

$$P_{k,s,h}^2 \approx \sum_{l=1}^{L} \alpha_{k,l} p_{k,s,\,h,l} \tag{B1}$$

$$Q_{k,s,h}^2 \approx \sum_{l=1}^{L} \beta_{k,l} q_{k,s,\,h,l} \tag{B2}$$

$$P_{k,s,h}^+ + P_{k,s,h}^- = \sum_{l=1}^{L} \alpha_{k,l} p_{k,s,\,h,l} \tag{B3}$$

$$Q_{k,s,h}^+ + Q_{k,s,h}^- = \beta_{k,l} q_{k,s,\,h,l} \tag{B4}$$

where $p_{k,s,h,l} \le P_k^{max}/L$; $q_{k,s,h,l} \le Q_k^{max}/L$; $\alpha_{k,l} = (2l-1)P_k^{max}/L$; and $\beta_{k,l} = (2l-1)Q_k^{max}/L$.

References

1. Papaefthymiou, G.; Dragoon, K. Towards 100% renewable energy systems: Uncapping power system flexibility. *Energy Policy* **2016**, *92*, 69–82. [CrossRef]
2. International Energy Agency (IEA). *Empowering Variable Renewables—Options for Flexible Electricity Systems*; OECD: Paris, France, 2008.
3. Santos, S.F.; Fitiwi, D.Z.; Cruz, M.R.M.; Cabrita, C.M.P.; Catalão, J.P.S. Impacts of optimal energy storage deployment and network reconfiguration on renewable integration level in distribution systems. *Appl. Energy* **2017**, *185*, 44–55. [CrossRef]
4. Brouwer, A.S.; van den Broek, M.; Seebregts, A.; Faaij, A. Impacts of large-scale Intermittent Renewable Energy Sources on electricity systems, and how these can be modeled. *Renew. Sustain. Energy Rev.* **2014**, *33*, 443–466. [CrossRef]
5. Ceaki, O.; Vatu, R.; Mancasi, M.; Porumb, R.; Seritan, G. Analysis of Electromagnetic Disturbances for Grid-Connected PV Plants. In Proceedings of the 2015 MEPS—International Conference Modern Electric Power Systems, Wroclaw, Poland, 6–9 July 2015.
6. Ceaki, O.; Vatu, R.; Golovanov, N.; Porumb, R.; Seritan, G. Analysis of the grid-connected PV plants behavior with FACTS influence. In Proceedings of the 49th International Universities Power Engineering Conference (UPEC), Cluj-Napoca, Romania, 2–5 September 2014.
7. Panagiotis, K.; Lambros, E. *Electricity Distribution*; Springer: New York, NY, USA, 2016.
8. Hossain, M.S.; Madlool, N.A.; Rahim, N.A.; Selvaraj, J.; Pandey, A.K.; Khan, A.F. Role of smart grid in renewable energy: An overview. *Renew. Sustain. Energy Rev.* **2016**, *60*, 1168–1184. [CrossRef]
9. Schachter, J.A.; Mancarella, P. A critical review of Real Options thinking for valuing investment flexibility in Smart Grids and low carbon energy systems. *Renew. Sustain. Energy Rev.* **2016**, *56*, 261–271. [CrossRef]
10. Yenginer, H.; Cetiz, C.; Dursun, E. A review of energy management systems for smart grids. In Proceedings of the 2015 3rd International Istanbul, Smart Grid Congress and Fair (ICSG), Istanbul, Turkey, 29–30 April 2015; pp. 1–4.
11. Kulkarni, S.N.; Shingare, P. A review on Smart Grid Architecture and Implementation Challenges. In Proceedings of the 2016 International Conference on Electrical, Electronics, and Optimization Techniques (ICEEOT), Chennai, India, 3–5 March 2016.
12. Zhou, X.; Cui, H.; Ma, Y.; Gao, Z. Research review on smart distribution grid. In Proceedings of the 2016 IEEE International Conference on Mechatronics and Automation (ICMA), Harbin, China, 7–10 August 2016; pp. 575–580.

13. Udgave, A.D.; Jadhav, H.T. A review on Distribution Network protection with penetration of Distributed Generation. In Proceedings of the 2015 IEEE 9th International Conference on Intelligent Systems and Control (ISCO), Coimbatore, India, 9–10 January 2015; pp. 1–4.

14. Bhimarasetti, R.T.; Kumar, A. A New Contribution to Distribution Load Flow Analysis for Radial and Mesh Distribution Systems. In Proceedings of the 2014 International Conference on Computational Intelligence and Communication Networks, Bhopal, India, 14–16 November 2014; pp. 1229–1236.

15. Arritt, R.F.; Dugan, R.C. Review of the Impacts of Distributed Generation on Distribution Protection. In Proceedings of the 2015 IEEE Rural Electric Power Conference, Asheville, NC, USA, 19–21 April 2015; pp. 69–74.

16. Tiwari, A.K.; Mohanty, S.R.; Singh, R.K. Review on protection issues with penetration of distributed generation in distribution system. In Proceedings of the 2014 International Electrical Engineering Congress (iEECON), Chonburi, Thailand, 19–21 March 2014; pp. 1–4.

17. Celli, G.; Pilo, F.; Pisano, G.; Cicoria, R.; Iaria, A. Meshed vs. radial MV distribution network in presence of large amount of DG. In Proceedings of the IEEE PES Power Systems Conference and Exposition, New York, NY, USA, 10–13 October 2004; pp. 1357–1362.

18. Yu, P.; Venkatesh, B.; Yazdani, A.; Singh, B.N. Optimal Location and Sizing of Fault Current Limiters in Mesh Networks Using Iterative Mixed Integer Nonlinear Programming. *IEEE Trans. Power Syst.* **2016**, *31*, 4776–4783. [CrossRef]

19. Zubo, R.H.A.; Mokryani, G.; Rajamani, H.-S.; Aghaei, J.; Niknam, T.; Pillai, P. Operation and planning of distribution networks with integration of renewable distributed generators considering uncertainties: A review. *Renew. Sustain. Energy Rev.* **2017**, *72*, 1177–1198. [CrossRef]

20. Alvarez-Herault, M.-C.; N'Doye, N.; Gandioli, C.; Hadjsaid, N.; Tixador, P. Meshed distribution network vs reinforcement to increase the distributed generation connection. *Sustain. Energy Grids Netw.* **2015**, *1*, 20–27. [CrossRef]

21. Davoudi, M.; Cecchi, V.; Agüero, J.R. Increasing penetration of distributed generation with meshed operation of distribution systems. In Proceedings of the North American Power Symposium (NAPS), Pullman, WA, USA, 7–9 September 2014; pp. 1–6.

22. Chalapathi, B.; Agrawal, D.; Murty, V.; Kumar, A. Optimal placement of Distribution Generation in weakly meshed Distribution Network for energy efficient operation. In Proceedings of the 2015 Conference on Power, Control, Communication and Computational Technologies for Sustainable Growth (PCCCTSG), Kurnool, India, 11–12 December 2015; pp. 150–155.

23. Ivic, D.; Macanovic, D.; Sosic, D.; Stefanov, P. Weakly meshed distribution networks with distributed generation—Power flow analysis using improved impedance matrix based algorithm. In Proceedings of the International Symposium on Industrial Electronics (INDEL), Banja Luka, Bosnia Herzegovina, 3–5 November 2016; pp. 1–6.

24. Yang, H.; Bae, T.; Kim, J.; Kim, Y.H. Load model technique for mesh-structured power distribution network. In Proceedings of the 2012 4th Asia Symposium on Quality Electronic Design (ASQED), Penang, Malaysia, 10–11 July 2012; pp. 219–222.

25. Yu, L.; Czarkowski, D.; de León, F.; Bury, W. A time sequence load-flow method for steady-state analysis in heavily meshed distribution network with DG. In Proceedings of the 2013 8th International Conference on Compatibility and Power Electronics (CPE), Ljubljana, Slovenia, 5–7 June 2013; pp. 25–30.

26. Fitiwi, D.Z.; Olmos, L.; Rivier, M.; de Cuadra, F.; Pérez-Arriaga, I.J. Finding a representative network losses model for large-scale transmission expansion planning with renewable energy sources. *Energy* **2016**, *101*, 343–358. [CrossRef]

27. Zhang, H.; Heydt, G.T.; Vittal, V.; Quintero, J. An Improved Network Model for Transmission Expansion Planning Considering Reactive Power and Network Losses. *IEEE Trans. Power Syst.* **2013**, *28*, 3471–3479. [CrossRef]

![energies logo] *energies*

MDPI

Article

Generation Expansion Planning Model for Integrated Energy System Considering Feasible Operation Region and Generation Efficiency of Combined Heat and Power

Woong Ko [1] and Jinho Kim [2],*

[1] Research Institute for Solar and Sustainable Energies, Gwangju Institute of Science and Technology, 123 Cheomdangwagi-ro, Buk-gu, Gwangju 61005, Korea; kwoong2001@gist.ac.kr
[2] School of Integrated Technology, Gwangju Institute of Science and Technology, 123 Cheomdangwagi-ro, Buk-gu, Gwangju 61005, Korea
* Correspondence: jeikim@gist.ac.kr; Tel./Fax: +82-62-715-5322

Received: 6 December 2018; Accepted: 11 January 2019; Published: 11 January 2019

Abstract: Integrated energy systems can provide a more efficient supply than individual systems by using resources such as cogeneration. To foster efficient management of these systems, the flexible operation of cogeneration resources should be considered for the generation expansion planning model to satisfy the varying demand of energy including heat and electricity, which are interdependent and present different seasonal characteristics. We propose an optimization model of the generation expansion planning for an integrated energy system considering the feasible operation region and efficiency of a combined heat and power (CHP) resource. The proposed model is formulated as a mixed integer linear programming problem to minimize the sum of the annualized cost of the integrated energy system. Then, we set linear constraints of energy resources and describe linearized constraints of a feasible operation region and a generation efficiency of the CHP resource for application to the problem. The effectiveness of the proposed optimization problem is verified through a case study comparing with results of a conventional optimization model that uses constant heat-to-power ratio and generation efficiency of the CHP resource. Furthermore, we evaluate planning schedules and total generation efficiency profiles of the CHP resource for the compared optimization models.

Keywords: integrated energy system; generation expansion planning; combined heat and power; feasible operation region; generation efficiency; mixed integer linear programming

---

1. Introduction

Modern energy systems tend to have a decentralized management based on energy interdependencies among different energy sources to increase operating efficiency [1]. Although such systems have different denominations, such as integrated energy systems, multi-energy systems and so on, these systems could be defined as integrated energy systems since they all aim to supply energy load from different carriers. The integrated energy systems are applied in various sectors to conform buildings, communities, and other energy demands [2,3]. Furthermore, these systems are utilized for introducing a large share of renewable energy into conventional power systems. In the Reference [4], an expansion planning method for the integrated energy system was proposed for minimizing the amount of curtailed energy in a Caribbean island, where renewable energy is the main source to supply electricity demand. In addition, an energy availability under stochastic nature of renewable energy improved when electrical and thermal generation were incorporated [5]. The operational dispatch strategies with multiple energy sources reduced negative effects of intermittent renewable energy on small energy systems [6,7]. Therefore, the integrated energy systems have the advantages of reducing the usage of fossil fuels and increasing the penetration rate of renewable energy.

A key component to construct the integrated energy systems is a combined heat and power (CHP) resource. This resource can simultaneously provide heat and electricity with a single source, such as gas and is environment-friendly because it reduces greenhouse gas emissions [8,9]. For efficient operation, other resources are utilized with the CHP resource in integrated energy systems [10,11]. Therefore, developing a methodology of generation expansion planning for these systems should thoroughly consider each of these resources.

Generation expansion planning models for integrated energy systems have been extensively studied. In one model of integrated energy systems called energy hub, this model aims to optimize multi-energy management considering three energy carriers, namely, electricity, heat and gas [12,13]. In addition, this model considers the interdependency among energy carriers for generation expansion planning. A sustainable framework for optimal energy hub design under unpredictable weather conditions and uncertainty of load demand was proposed in Reference [14]. This framework used the Benders decomposition to minimize planning and operation costs of an energy hub with various CHP resources and other dispatchable and non-dispatchable distributed energy resources. Likewise, power system reliability indices, such as loss of load probability and expected energy not served, have been applied for optimal energy hub design [15,16]. Furthermore, the environmental impact regarding aspects such as carbon emissions has been considered in recent research on the energy hub model [17]. Microgrids conform another model of the integrated energy systems. In Reference [18], an integrated energy microgrid planning considers the demand of heat, electricity and cooling from energy interdependent system. Additionally, this planning method aimed to improve energy supply for multiple regions.

The variations of heat and electricity demand should be considered in the generation expansion planning for the integrated energy systems because planning horizons usually range from a single to multiple years. For example, heat demand is usually low during summer but high during winter. As part of the system, the seasonal variations of energy loads also influence the operation of the CHP resources, whose amount of heat and electricity generation is determined by a heat-to-power ratio. This ratio close to 1 indicates a high overall generation efficiency of the CHP resource [19]. However, during summer, the wasted heat from the CHP resource would increase if the heat-to-power ratio is maintained at constant value. Many studies have been focused on addressing seasonal variations. In Reference [20,21], the planning strategy for optimal usage of seasonal thermal energy storage was proposed by minimizing the fuel cost incurred by the operation of the heat-only boiler of the CHP plant. In Reference [22], an absorption chiller using waste heat was utilized to improve the overall generation efficiency of the CHP resource. Besides utilizing other facilities, a feasible operation region of the CHP resources should be considered to address seasonal variations [23]. The feasible operation region allows the CHP resources to have varying values of heat-to-power ratio and has been used for flexible heat and electricity generation. In Reference [24], optimized production scheduling of the CHP resource aimed to reduce emissions by using this region. In Reference [25], the CHP resources characterized by this region and a thermal energy storage were used for leading to larger revenue from power sales. These studies mainly used this region as operation constraints of their production optimization because the seasonal variation of heat demand is higher than that of electricity demand.

Nevertheless, research on the generation expansion planning for the integrated energy systems rarely considers the seasonal characteristics of loads and the feasible operation region of the CHP resource. Although the planning horizon is usually enough to consider seasonal characteristics of loads, most of generation expansion planning models have been simply focused on the installation of energy resources with a constant heat-to-power ratio of the CHP resource [26,27]. These omissions in the generation expansion planning model can result in either over or underestimated operation costs. To overcome these modeling limitations, we propose a method of generation expansion planning model for an integrated energy system considering the feasible operation region and generation efficiency of the CHP resource. This paper addresses the following aspects:

- Optimization problem of generation expansion planning for an integrated energy system is modeled as a mixed integer linear programming (MILP) problem considering energy resources, including the CHP resource, fuel-based generators and energy storage resources.
- Feasible operation region and a generation efficiency function of the CHP resource are modeled as linear constraints of the MILP problem.
- To validate the proposed method, the conventional optimization model using constant heat-to-power ratio and generation efficiency of the CHP resource is compared to the proposed optimization model in a case study.

The rest of this paper is organized as follows. Section 2 describes generation expansion planning model for integrated energy systems along with its objective function and constraints. We then introduce the optimization model considering the feasible operation region and generation efficiency function of the CHP resource in Section 3. A comparison between the conventional and the proposed expansion planning is detailed through a case study in Section 4 and we draw conclusions in Section 5.

2. Generation Expansion Planning Model for Integrated Energy Systems

This section provides an overview of the integrated energy system and optimization model for expansion planning using MILP.

2.1. Integrated Energy System Model

We use an integrated energy system model from a previous study [28]. The system is assumed to be self-sufficient and its model is depicted in Figure 1. In the model, the CHP resource as well as heat and electricity generation resources are depicted with unidirectional units. In contrast, thermal and electrical energy storage are configured as bidirectional units given their ability for charging and discharging.

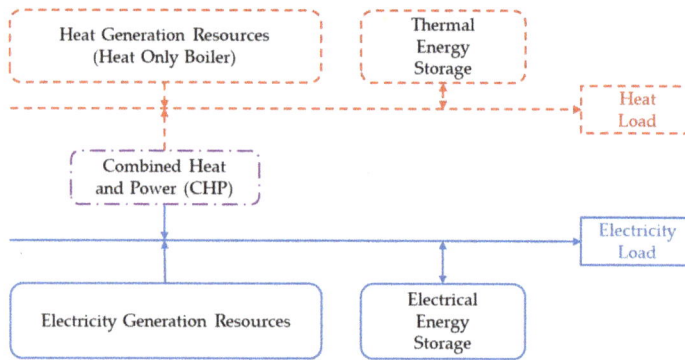

Figure 1. Integrated energy system model.

2.2. Objective Function

The proposed generation expansion planning model for an integrated energy system aims to minimize the sum of the annualized costs, including initial investment and operation costs of the CHP resource as well as heat and electrical energy resources over planning horizon N_Y. Hence, the objective function is defined as

$$\text{Minimize} \sum_{y=1}^{N_Y} \left((1 + \gamma_d)^{-y} \cdot \left(COST_e^y + COST_h^y + COST_{CHP}^y \right) \right), \tag{1}$$

considering costs

$$\text{COST}_e^y = \sum_{i=1}^{N_{ER}} \left(\left(CRF_e^i \cdot CC_e^i + FOMC_e^i \right) \cdot \sum_{c=1}^{N_C} \left(C_e^{i,c} \cdot v_e^{i,c,y} \right) + \left(FC_e^i + VOMC_e^i \right) \cdot \sum_{t=1}^{N_T} F_e^{i,t,y} \right), \quad (2)$$

$$\text{COST}_h^y = \sum_{j=1}^{N_{HR}} \left(\left(CRF_h^j \cdot CC_h^j + FOMC_h^j \right) \cdot \sum_{c=1}^{N_C} \left(C_h^{j,c} \cdot v_h^{j,c,y} \right) + \left(FC_h^j + VOMC_h^j \right) \cdot \sum_{t=1}^{N_T} F_h^{j,t,y} \right), \quad (3)$$

$$\text{COST}_{CHP}^y = \sum_{ch=1}^{N_{CHP}} \left(\left(CRF_{CHP}^{ch} \cdot CC_{CHP}^{ch} + FOMC_{CHP}^{ch} \right) \cdot \sum_{c=1}^{N_C} \left(C_{CHP}^{ch,c} \cdot v_{CHP}^{ch,c,y} \right) \right)$$
$$+ \sum_{ch=1}^{N_{CHP}} \left(\left(FC_{CHP}^{ch} + VOMC_{CHP}^{ch} \right) \cdot \sum_{t=1}^{N_T} F_{CHP}^{ch,t,y} \right) \quad \forall y, \quad (4)$$

where

$$CRF_e^i = \frac{\gamma_d (1+\gamma_d)^{LT_e^i}}{(1+\gamma_d)^{LT_e^i} - 1}, \quad CRF_h^j = \frac{\gamma_d (1+\gamma_d)^{LT_h^j}}{(1+\gamma_d)^{LT_h^j} - 1}, \quad CRF_{CHP}^{ch} = \frac{\gamma_d (1+\gamma_d)^{LT_{CHP}^{ch}}}{(1+\gamma_d)^{LT_{CHP}^{ch}} - 1} \forall i, \forall j, \forall ch, \quad (5)$$

are the capital recovery factors from electrical and heat energy resources and CHP resources. COST_e^y, COST_h^y and COST_{CHP}^y are the respective total costs by energy resources for planning year y. Each cost comprises the annual fixed costs (including capital costs and fixed operation and maintenance costs related to the resource capacity) and the annual variable costs (including fuel costs and variable operation and maintenance costs related to fuel consumption). Equations (2)–(4) express the respective sum of annualized fixed and variable costs of resources. The annual fixed costs per unit capacity consist of capital recovery factors (CRF_e^i, CRF_h^j and CRF_{CHP}^{ch}), overnight capital costs (CC_e^i, CC_h^j and CC_{CHP}^{ch}) and fixed operation and maintenance costs ($FOMC_e^i$, $FOMC_h^j$ and $FOMC_{CHP}^{ch}$). The fixed costs are determined by the capacity of a candidate unit ($C_e^{i,c}$, $C_h^{j,c}$ and $C_{CHP}^{ch,c}$) of being selected with corresponding decision variable ($v_e^{i,c,y}$, $v_h^{j,c,y}$ and $v_{CHP}^{ch,c,y}$). Likewise, the annual variable costs per unit consist of fuel costs (FC_e^i, FC_h^j and FC_{CHP}^{ch}) and variable operation and maintenance costs ($VOMC_e^i$, $VOMC_h^j$ and $VOMC_{CHP}^{ch}$). The variable costs are determined by the fuel consumption per hour ($F_e^{i,t,y}$, $F_h^{j,t,y}$ and $F_{CHP}^{ch,t,y}$).

2.3. Constraints for Heat and Electrical Energy Resources

The fuel consumption of heat and electrical energy resources depends on their generation efficiency and power output:

$$F_e^{i,t,y} = \frac{P_e^{i,t,y}}{\eta_e^i}, \quad F_h^{j,t,y} = \frac{P_h^{j,t,y}}{\eta_h^j} \forall j, \forall t, \forall y. \quad (6)$$

The output of heat and electricity is limited by the capacity of the selected resource and its minimum and maximum generation levels are given by

$$\gamma_{e,min}^i \cdot \sum_{c=1}^{N_C} \left(C_e^{i,c} \cdot v_e^{i,c,y} \right) \leq P_e^{i,t,y} \leq \gamma_{e,max}^i \cdot \sum_{c=1}^{N_C} \left(C_e^{i,c} \cdot v_e^{i,c,y} \right) \forall j, \forall t, \forall y,$$

$$\gamma_{h,min}^j \cdot \sum_{c=1}^{N_C} \left(C_h^{j,c} \cdot v_h^{j,c,y} \right) \leq P_h^{j,t,y} \leq \gamma_{h,max}^j \cdot \sum_{c=1}^{N_C} \left(C_h^{j,c} \cdot v_h^{j,c,y} \right) \forall j, \forall t, \forall y. \quad (7)$$

2.4. Constraints for Electrical and Thermal Energy Storage Resources

The energy storage resources can charge and discharge energy within their state of charge (SOC) limits. Hence, stored energy is limited by the minimum and maximum SOC levels:

$$
SOC_{EES,min} \cdot \sum_{i=1}^{N_{ER}} \left(\rho_{EES}^i \cdot \sum_{c=1}^{N_C} \left(C_e^{i,c} \cdot v_e^{i,c,y} \right) \right) \le E_{EES}^{t,y} \le SOC_{EES,max} \cdot \sum_{i=1}^{N_{ER}} \left(\rho_{EES}^i \cdot \sum_{c=1}^{N_C} \left(C_e^{i,c} \cdot v_e^{i,c,y} \right) \right),
$$
$$
SOC_{TES,min} \cdot \sum_{j=1}^{N_{HR}} \left(\rho_{TES}^j \cdot \sum_{c=1}^{N_C} \left(C_h^{j,c} \cdot v_h^{j,c,y} \right) \right) \le E_{TES}^{t,y} \le SOC_{TES,max} \cdot \sum_{j=1}^{N_{HR}} \left(\rho_{TES}^j \cdot \sum_{c=1}^{N_C} \left(C_h^{j,c} \cdot v_h^{j,c,y} \right) \right).
$$

$$(8)$$

The amount of stored energy decreases or increases during discharging or charging, respectively, except for the initially stored energy, as described by

$$
E_{EES}^{t,y} =
\begin{cases}
E_{EES}^{0,y}, & \text{if } t = 1 \\
E_{EES}^{t-1,y} - \sum\limits_{i=1}^{N_{ER}} \left(\rho_{EES}^i \cdot P_e^{i,t-1,y} / \eta_{EES,eff} \right) + P_{ch,e}^{t-1,y} \eta_{EES,eff}, & \text{otherwise}
\end{cases}
\forall t, \forall y,
$$

$$
E_{TES}^{t,y} =
\begin{cases}
E_{TES}^{0,y}, & \text{if } t = 1 \\
E_{TES}^{t-1,y} - \sum\limits_{j=1}^{N_{HR}} \left(\rho_{TES}^j \cdot P_h^{j,t-1,y} / \eta_{TES,eff} \right) + P_{ch,h}^{t-1,y} \eta_{TES,eff}, & \text{otherwise}
\end{cases}
\forall t, \forall y.
$$

$$(9)$$

The amount of discharging or charging energy is limited by the capacity of energy storage resources, which can only operate in one mode at a given time. These storage characteristics can be described as

$$
\sum_{i=1}^{N_{ER}} \left(\rho_{EES}^i \cdot P_e^{i,t,y} \right) \le \gamma_{EES,dch} \cdot \left(1 - v_{ch,e}^{t,y} \right) \cdot \sum_{i=1}^{N_{ER}} \left(\rho_{EES}^i \cdot \sum_{c=1}^{N_C} \left(C_e^{i,c} \cdot v_e^{i,c,y} \right) \right) \forall t, \forall y,
$$
$$
\sum_{j=1}^{N_{HR}} \left(\rho_{TES}^j \cdot P_h^{j,t,y} \right) \le \gamma_{TES,dch} \cdot \left(1 - v_{ch,h}^{t,y} \right) \cdot \sum_{j=1}^{N_{HR}} \left(\rho_{TES}^j \cdot \sum_{c=1}^{N_C} \left(C_h^{j,c} \cdot v_h^{j,c,y} \right) \right) \forall t, \forall y,
$$

$$(10)$$

$$
P_{ch,e}^{t,y} \le \gamma_{EES,ch} \cdot v_{ch,e}^{t,y} \cdot \sum_{i=1}^{N_{ER}} \left(\rho_{EES}^i \cdot \sum_{c=1}^{N_C} \left(C_e^{i,c} \cdot v_e^{i,c,y} \right) \right) \forall t, \forall y,
$$
$$
P_{ch,h}^{t,y} \le \gamma_{TES,ch} \cdot v_{ch,h}^{t,y} \cdot \sum_{j=1}^{N_{HR}} \left(\rho_{TES}^j \cdot \sum_{c=1}^{N_C} \left(C_h^{j,c} \cdot v_h^{j,c,y} \right) \right) \forall t, \forall y,
$$

$$(11)$$

$$
v_{ch,e}^{t,y} \le 1 \, \forall t, \forall y,
$$
$$
v_{ch,h}^{t,y} \le 1 \, \forall t, \forall y.
$$

$$(12)$$

Note that linearization is required for MILP, because Equations (10) and (11) are nonlinear constraints. This linearization process is detailed in Appendix A.

2.5. Energy Balance Constraints

The heat and electrical load and charging load are supplied every hour by all energy resources, as described by

$$
\sum_{i=1}^{N_{ER}} P_e^{i,t,y} + \sum_{ch=1}^{N_{CHP}} P_{CHP,e}^{ch,t,y} = ld_e^{t,y} + P_{ch,e}^{t,y} \, \forall t, \forall y,
$$

$$
\sum_{j=1}^{N_{HR}} P_h^{j,t,y} + \sum_{ch=1}^{N_{CHP}} P_{CHP,h}^{ch,t,y} = ld_h^{t,y} + P_{ch,h}^{t,y} \, \forall t, \forall y.
$$

$$(13)$$

3. Feasible Operation Region and Generation Efficiency of CHP Resource

This section describes the feasible operation region and the generation efficiency functions of the CHP resource. In addition, these are modeled with optimization constraints for using MILP.

3.1. Feasible Operation Region for CHP Resource

Heat and electricity generated from the CHP resource depend on the feasible operation region [29,30], which is depicted by polyhedrons as shown in Figure 2.

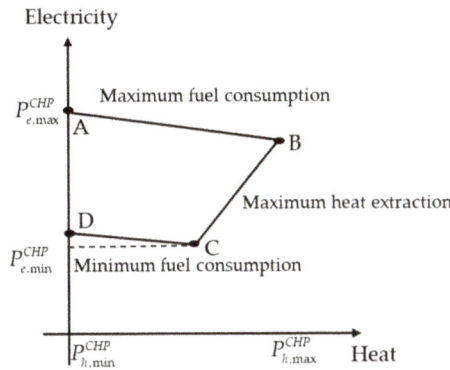

Figure 2. Feasible Operation Region of combined heat and power (CHP).

The polyhedral region consists of four CHP operating points. The points A and B represent the maximum electricity and heat generation, respectively and their joining segment defines the maximum fuel consumption. The points C and D represent the minimum electricity generation and heat generation, respectively and their joining segment defines the minimum fuel consumption. The segments between the points A and D and between the points B and C define the operation range of only electricity generation and maximum heat extraction, respectively.

Heat and electricity generation belong to the region within the abovementioned segments and are defined by Equations (14)–(17). The following condition must be satisfied for the CHP resource to generate heat and electrical energy when the resource is committed:

$$0 \leq P_{CHP,e}^{ch,t,y} \leq P_{CHP,e}^{ch,A} \cdot v_{CHP}^{ch,y} \ \forall t, \ \forall y,$$
$$0 \leq P_{CHP,h}^{ch,t,y} \leq P_{CHP,h}^{ch,B} \cdot v_{CHP}^{ch,y} \ \forall t, \ \forall y,$$

(14)

where $P_{CHP,e}^{ch,A}$ and $P_{CHP,h}^{ch,B}$ are the maximum operating points of electricity and heat generation, respectively. The operating condition within the segment of maximum fuel consumption is defined as

$$P_{CHP,e}^{ch,t,y} - P_{CHP,e}^{ch,A} - \frac{\left(P_{CHP,e}^{ch,A} - P_{CHP,e}^{ch,B} \right)}{\left(P_{CHP,h}^{ch,A} - P_{CHP,h}^{ch,B} \right)} \left(P_{CHP,h}^{ch,t,y} - P_{CHP,h}^{ch,A} \right) \leq 0 \ \forall t, \ \forall y.$$

(15)

Other operating conditions within the segments of maximum heat extraction and minimum fuel consumption are respectively defined as

$$P_{CHP,e}^{ch,t,y} - P_{CHP,e}^{ch,B} - \frac{\left(P_{CHP,e}^{ch,B} - P_{CHP,e}^{ch,C} \right)}{\left(P_{CHP,h}^{ch,B} - P_{CHP,h}^{ch,C} \right)} \left(P_{CHP,h}^{ch,t,y} - P_{CHP,h}^{ch,B} \right) \geq -\left(1 - v_{CHP}^{ch,y} \right) \cdot Z \ \forall t, \ \forall y,$$

(16)

$$P_{CHP,e}^{ch,t,y} - P_{CHP,e}^{ch,C} - \frac{\left(P_{CHP,e}^{ch,C} - P_{CHP,e}^{ch,D}\right)}{\left(P_{CHP,h}^{ch,C} - P_{CHP,h}^{ch,D}\right)}\left(P_{CHP,h}^{ch,t,y} - P_{CHP,h}^{ch,D}\right) \geq -\left(1 - v_{CHP}^{ch,y}\right) \cdot Z \; \forall t, \; \forall y, \qquad (17)$$

where $P_{CHP,e}^{ch,C}$ and $P_{CHP,h}^{ch,D}$ are the minimum operating points of electricity and heat generation, respectively and Z is a sufficiently large constant close to positive infinity. These equations of the operating segments are greater than negative infinity or zero, $\left(-\left(1 - v_{CHP}^{ch,y}\right) \cdot Z\right)$. These inequalities without negative infinity are not satisfied if the CHP resource is not committed, because these regions do not contain a zero operating point of electricity generation.

3.2. Efficiency Functions of CHP Resource

We assume that the generation efficiency of the CHP resource is varied with the generated heat or electricity over the feasible operation region based on the function of loading level of the CHP resource [31]. The efficiency functions for the CHP resource are also assumed to be modeled by discrete functions, which are required for MILP and illustrate the relationship between generation efficiency and output, as shown in Figure 3.

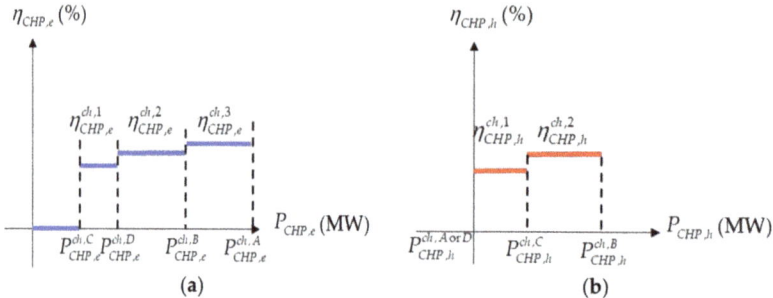

Figure 3. Efficiency functions for CHP resource: (**a**) Electricity generation efficiency function; (**b**) heat generation efficiency function.

The efficiency functions for the CHP resource are assumed to be modeled by the mean value of the generation efficiency over segments on the feasible operation region. Therefore, the number of boundary segments of these functions are 3 and 2, respectively.

Variables for selecting efficiency segments are determined considering the generation efficiency varying with the generation of the CHP resource as follows:

$$v_{CHP,eff,e}^{ch,be,t,y} \in \{0,1\} \forall ch, \; \forall be, \; \forall t, \; \forall y,$$
$$v_{CHP,eff,h}^{ch,bh,t,y} \in \{0,1\} \forall ch, \; \forall bh, \; \forall t, \; \forall y, \qquad (18)$$

where be and bh are indices of the boundary segments of electricity and heat generation efficiency function, respectively. Each of these variables is equal to 1 if the corresponding CHP resource generates electrical energy determined in the specific efficiency segment and it is equal to 0 otherwise. Examples of segment selection with these variables are illustrated in Figure 4.

The generation output of the CHP resource should be defined in the boundary segments of the efficiency function to also guarantee that the output is within the feasible operation region. The heat and electricity generation is defined as

$$
\begin{cases}
P_{CHP,e}^{ch,t,y} \leq \left(1 - v_{CHP,eff,e}^{ch,be,t,y}\right) \cdot Z + v_{CHP,eff,e}^{ch,be,t,y} \cdot P_{CHP,e}^{ch,be+1} \\
P_{CHP,e}^{ch,t,y} \geq -\left(1 - v_{CHP,eff,e}^{ch,be,t,y}\right) \cdot Z + v_{CHP,eff,e}^{ch,be,t,y} \cdot P_{CHP,e}^{ch,be}
\end{cases} \quad \forall ch, \forall be, \forall t, \forall y,
$$
$$
\begin{cases}
P_{CHP,h}^{ch,t,y} \leq \left(1 - v_{CHP,eff,h}^{ch,bh,t,y}\right) \cdot Z + v_{CHP,eff,h}^{ch,bh,t,y} \cdot P_{CHP,h}^{ch,bh+1} \\
P_{CHP,h}^{ch,t,y} \geq -\left(1 - v_{CHP,eff,h}^{ch,bh,t,y}\right) \cdot Z + v_{CHP,eff,h}^{ch,bh,t,y} \cdot P_{CHP,h}^{ch,bh}
\end{cases} \quad \forall ch, \forall bh, \forall t, \forall y.
$$

$$(19)$$

For example, if the electricity generation of the committed CHP resource is within the first boundary segment of Figure 4a, the first boundary segment selecting variable is 1. Then, according to Equation (19), the electricity generation is bigger than $P_{CHP,e}^{ch,C}$ and lower than $P_{CHP,e}^{ch,D}$. In other boundary segments, the electricity generation is bigger than negative infinity and lower than positive infinity since other selecting variables are 0. The additional condition is required since the boundary segment can only be selected once or not at all:

$$
\sum_{be=1}^{N_{BE}} v_{CHP,eff,e}^{ch,be,t,y} = v_{CHP}^{ch,y} \quad \forall ch, \forall t, \forall y,
$$
$$
\sum_{bh=1}^{N_{BH}} v_{CHP,eff,h}^{ch,bh,t,y} = v_{CHP}^{ch,y} \quad \forall ch, \forall t, \forall y.
$$

$$(20)$$

With these conditions, the variables for selecting boundary segments can be decided according to the generation output.

Finally, the fuel consumption of the CHP resource is defined as the generation output and efficiency determined by the abovementioned procedures:

$$
F_{CHP,e}^{ch,t,y} = \sum_{be=1}^{N_{BE}} \left(\frac{P_{CHP,e}^{ch,t,y}}{\eta_{CHP,e}^{ch,be}} \cdot v_{CHP,eff,e}^{ch,be,t,y} \right) \quad \forall ch, \forall t, \forall y,
$$
$$
F_{CHP,h}^{ch,t,y} = \sum_{bh=1}^{N_{BH}} \left(\frac{P_{CHP,h}^{ch,t,y}}{\eta_{CHP,h}^{ch,bh}} \cdot v_{CHP,eff,h}^{ch,bh,t,y} \right) \quad \forall ch, \forall t, \forall y.
$$

$$(21)$$

Linearization is required because the constraints on fuel consumption of CHP resources correspond to the product of variables. This linearization process is detailed in Appendix B.

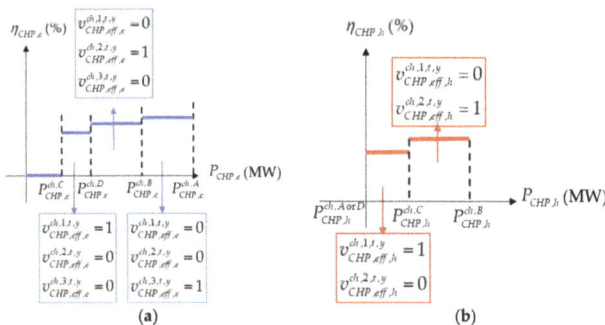

Figure 4. Examples of boundary segment selection for efficiency function: (**a**) Electricity generation efficiency function; (**b**) heat generation efficiency function.

4. Case Study and Discussion

We evaluated the effectiveness of the proposed method by comparing with the planning results using conventional generation expansion planning model.

4.1. Simulation Setup

We applied the proposed optimization model to a comprehensive integrated energy system with peak electricity demand of 1213.2 MW and peak heat demand of 956.8 MW [32,33]. The key parameters of the optimization model are listed in Table 1.

Table 1. Key parameters of generation expansion planning model.

Parameter	Value
Planning horizon (Year)	5
Planning horizon (hours in a year)	288
Interest rate (%)	3.91
Demand growth rate (%)	2.5

The planning horizon of 5 years considers 288 h (24 h × 12 months) per year instead of the total 8760 h to reduce the computational burden. The annual interest rate and demand growth rate are considered according to [34]. We considered the profiles of peak and mean load per month to generate the 288-h load profiles. The complete 8760-h load profiles were categorized by hour and month. For generating the peak load profiles, single-day load profiles containing the peak load per month were extracted from the 8760-h load profiles. To generate the mean load profiles, we calculated the average load every hour per month. Then, we collected the peak and mean load profiles considering the demand growth rate for 5 years to obtain the profiles shown in Figure 5.

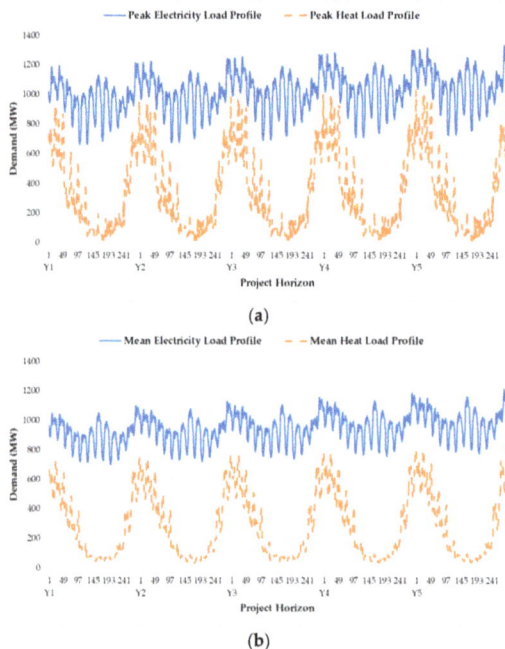

(a)

(b)

Figure 5. Load profiles over 288 hours in each of the 5 years (Y1–Y5): (**a**) Peak load profiles; (**b**) mean load profiles.

The cost and size data of the energy resources are listed in Appendix C. With the cost data, the operation data including the efficiency and operating point data are required. We used the operation data from the actual CHP resource technology listed in Table 2 [29,35,36]. The operating points in this table were designed based on the feasible operation region shown in Figure 2. In addition, operating points B and C of heat generation are equal to 0.92 times the electricity generation at operating points B and C, respectively, because the assumed CHP was designed with an average heat-to-power ratio of 0.92.

Table 2. Operating points and generation efficiency of the CHP resource.

Candidate Size of CHP Resource		1200 MW		1000 MW		800 MW	
	Symbol of Operating Point or Region	Electricity Generation	Heat Generation	Electricity Generation	Heat Generation	Electricity Generation	Heat Generation
Output (MW)	A	1380	0	1150	0	920	0
	B	1200	1104	1000	920	800	736
	C	480	331.2	400	276	320	220.8
	D	360	0	300	0	240	0
Generation Efficiency (%)	A–B	38	-	36	-	34	-
	B–C	30	42	28	44	26	46
	C–D	22	21	20	22	18	23

Apart from CHP, other energy resources were assumed to have a constant generation efficiency. We also used the minimum and maximum generation limits of electrical and heat energy resources listed in Table 3 and the operation conditions for energy storage listed in Table 4.

Table 3. Operating data of electrical and heat energy resources.

Resource Type	Unit Name	Generation Efficiency (%)	Minimum Generation Limit (%)	Maximum Generation Limit (%)
Fuel-based Power Generator	DG1	40	20	90
	DG2	30	20	90
Heat Only Boiler	HOB1	70	5	100

Table 4. Operating data of energy storage resources.

Resource Type	Unit Name	Minimum State of Charge (%)	Maximum State of Charge (%)	Maximum Generation Limit (%)	Maximum Charging/Discharging Rate (%)	Turn Around Efficiency (%)
Electrical Energy Storage	EES	10	100	100	50/50	90
Thermal Energy Storage	TES	10	100	100	50/50	90

4.2. Simulation Results

We solved the proposed MILP optimization problem using intlinprog with Gurobi optimization in MATLAB (Mathworks, Inc., Natick, MA, USA). The dual simplex algorithm terminated after reaching a 0.1% duality gap. We compared the planning results of the proposed method and those of conventional method using constant heat-to-power ratio and generation efficiency of the CHP resource for supplying the peak and mean loads from the profiles shown in Figure 5. The heat-to-power ratio was assumed to be 0.92 and the generation efficiency was assumed to be 80% from the sum of the first boundary segments of electrical and heat generation efficiency in Table 2. The maximum electricity and heat generation of the CHP resource were assumed to be point A of electricity and heat generation in Table 2. We classified optimization models applied to expansion planning as detailed in Table 5.

Table 5. Classification of optimization models.

Model	Description
A	Optimization applying constant heat-to-power ratio and generation efficiency of CHP
B	Proposed optimization

4.2.1. Peak Load Supply

The cost results by optimization models for peak load supply are listed in Table 6. In addition, the percent variation from the costs of model A to those of model B are listed in the table. Most of the costs including the total cost obtained from model A were higher than those obtained from model B. Although the costs of the CHP resource in models A and B were very similar, those of electrical and heat energy resources in model A are at least 78% higher than those in model B. Consequently, the cost results suggest the number of built electrical and heat resources in model A is greater than that in model B.

Table 6. Cost results by optimization models for peak load supply.

Costs ($)		Model A	Model B (Proposed Model)	Percent Variance $((A - B)/A \times 100)$ (%)
Total Cost		8.79×10^8	7.22×10^8	17.9
Costs of CHP Resource	Total Fixed Cost	5.39×10^8	5.39×10^8	0
	Total Variable Cost	1.43×10^8	1.57×10^8	−9.8
Costs of Electrical Energy Resources	Total Fixed Cost	1.25×10^8	1.44×10^7	88.5
	Total Variable Cost	4.31×10^7	6.35×10^6	85.3
Costs of Heat Energy Resources	Total Fixed Cost	2.65×10^7	5.84×10^6	78.0
	Total Variable Cost	2.10×10^6	2.11×10^4	99.0

The difference of the generation expansion planning results using models A and B is depicted in Figures 6 and 7. Figure 6 shows the planning schedules for supplying peak electricity load, whereas Figure 7 shows those for supplying peak heat load. The total capacity of electrical energy resources in the planning obtained from model A was much larger than that from model B as shown in Figure 6a, because electricity generation of the CHP resource in model A was limited by the constant heat-to-power ratio. In model A, even a high electricity demand retrieved a low amount of electricity generation of the CHP resource when heat generation was low. On the other hand, the total capacity using model B did not exceed 15% of the peak load per year as shown in Figure 6b. The electricity generation of the CHP resource in model B was not relatively limited by the heat-to-power ratio as in model A, because this resource can flexibly generate heat and electricity on the feasible operation region. Therefore, the difference between planning schedules using models A and B was considerable, especially during year Y3, where all the electricity generation resources were installed in model A for supplying the electricity peak load, whereas electricity generation resource DG1 was not installed in model B. This difference was notably reflected in the costs obtained from the models.

The planning schedules for supplying heat peak load were very similar to those for supplying electricity peak load. Heat generation of the CHP resource can supply enough heat load until planning year Y2, as the heat peak load was much below its capacity. In model B, energy storage resource TES was installed on planning year Y3 for utilizing surplus heat, whereas in model A, boiler HOB1 was installed from planning year Y4 for supplying heat load through the CHP resource.

The total generation efficiency profiles of the CHP resource by the optimization models for supplying peak load are depicted in Figure 8. The profile of model A corresponds to the horizontal line, whereas the profile of model B varies along the project horizon. Overall, the total generation efficiency of the CHP resource using model A was higher than that using model B. Hence, the total variable cost of the CHP resource using model B was mostly higher than that using model A. Using model B, the total generation efficiency can be inferred from periods in the project horizon. For instance, the total generation efficiency in the period from 1 to 48 (i.e., from January to February) was mostly the highest because the CHP resource should supply a large heat demand during winter. In contrast, the total generation efficiency in the period from 144 to 240 (i.e., from June to October) was mostly the lowest because heat demand was low. From planning year Y3, the total generation efficiency reached the lowest value during summer because the operation of energy storage resource TES substituted the small amount of heat generation required from the CHP resource.

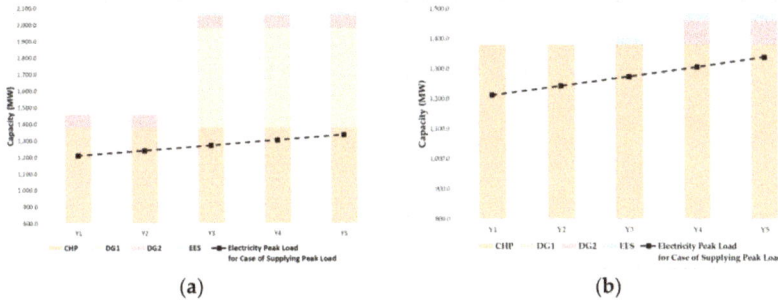

Figure 6. Planning schedules for supplying electricity peak load (**a**) using model A; (**b**) using model B.

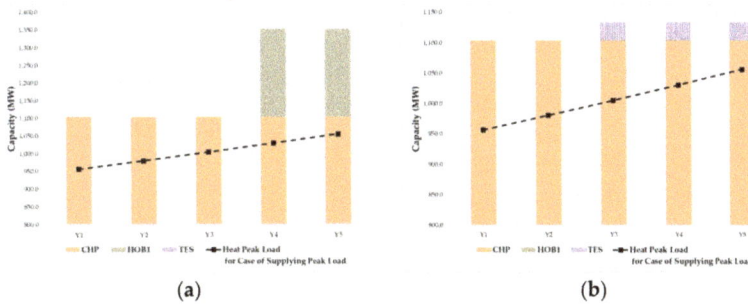

Figure 7. Planning schedules for supplying heat peak load (**a**) using model A; (**b**) using model B.

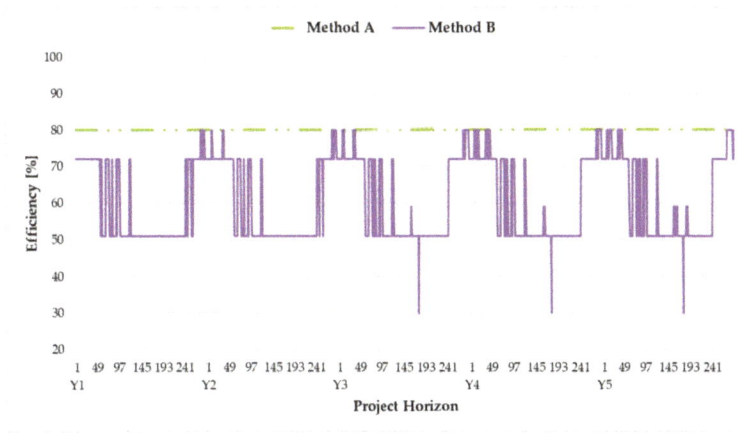

Figure 8. Total generation efficiency profiles of CHP by optimization models for peak load supply.

4.2.2. Mean Load Supply

The cost results by optimization models for mean load supply are listed in Table 7. Similar to the case of supplying peak load, the costs including the total cost obtained from model A were higher than those obtained from model B. Unlike the case of supplying peak load, the percent variance of the total cost between models A and B was small given the smaller electricity and heat demand in the mean load profile than in the peak load profile. Although the percent variance of costs of CHP resources and heat energy resources was small or zero, that of electrical energy resources was 100%. Overall, most of the heat and electricity mean load can be supplied by the CHP resource.

Table 7. Cost results by optimization models for mean load supply.

Costs ($)		Model A	Model B (Proposed Model)	Percent Variance ((A − B)/A×100) (%)
Total Cost		7.14×10^8	6.87×10^8	3.78
Costs of CHP Resources	Total Fixed Cost	5.39×10^8	5.39×10^8	0
	Total Variable Cost	1.68×10^8	1.48×10^8	1.19
Costs of Electrical Energy Resources	Total Fixed Cost	5.00×10^6	0	100
	Total Variable Cost	3.03×10^6	0	100
Costs of Heat Energy Resources	Total Fixed Cost	0	0	-
	Total Variable Cost	0	0	-

The difference of the generation expansion planning results using models A and B is depicted in Figures 9 and 10. Unlike the planning schedules in Figure 6, the CHP resource can cover the electricity load except for planning year Y5 using model A, as shown in Figure 9. Electrical energy resource DG2 should be installed to compensate the electricity shortage due to the constant heat-to-power ratio of the CHP resource when using model A. The expansion planning results for heat energy resources were equal regardless of the model, as shown in Figure 10. Only the CHP resource supplied the heat load without any additional resources.

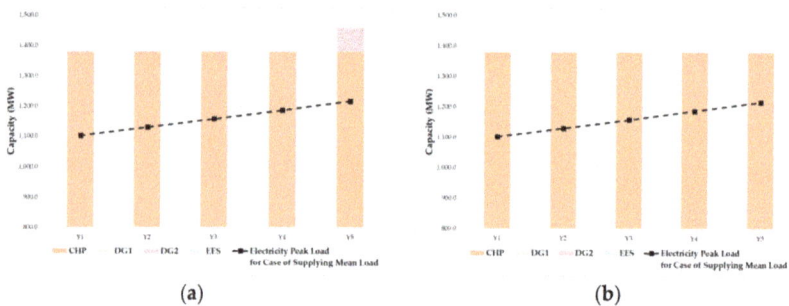

(a)   (b)

Figure 9. Planning schedules for supplying electric mean load (a) using model A; (b) using model B.

The total generation efficiency profiles of the CHP resource by the optimization models for mean load supply are depicted in Figure 11. Like in Figure 8, the total generation efficiency of the CHP resource using model A was higher than that using model B but for mean load, the profile using model B was more regular than that profile of supplying peak load, being between 50% and 80%. The profile using model B did not fluctuate greatly because CHP supplied most of the electricity and heat load.

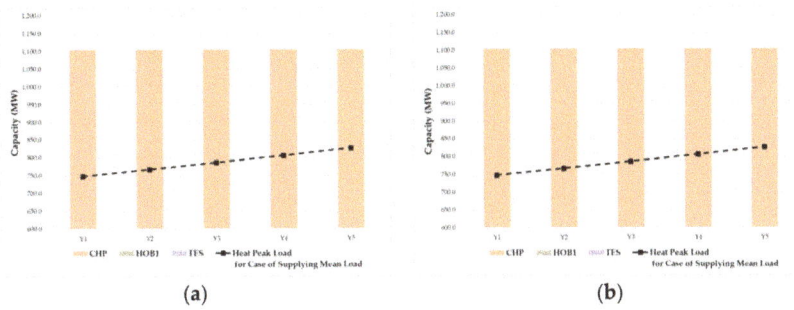

Figure 10. Planning schedules for supplying heat mean load (**a**) using model A; (**b**) using model B.

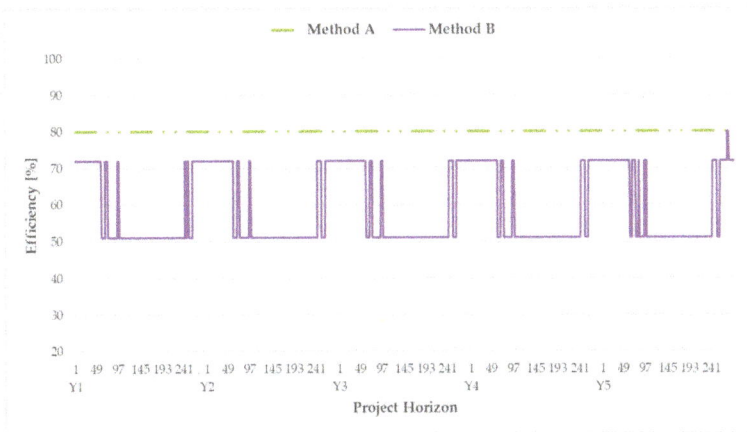

Figure 11. Total generation efficiency profiles of CHP by optimization models for mean load supply.

4.3. Discussions

The proposed optimization model could be utilized for either planning or managing energy resources in the integrated energy systems. Unlike conventional expansion planning models using constant heat-to-power ratio and generation efficiency, the proposed model could provide a cost-effective expansion plan as verified in the results. Moreover, the obtained results would be utilized as indicators for seasonal changes in operation of CHP resources. Although the generation efficiency of CHP resources should be thoroughly considered in energy expansion planning, most of the available studies and policies have overlooked details about the generation efficiency regarding seasonal changes. Therefore, the proposed model can provide more reliable scenarios for the integrated energy system planning under conditions that resemble the actual operation of CHP resources.

5. Conclusions

We propose a method of generation expansion planning for an integrated energy system. Although the objective function and constraints of the conventional heat and electrical energy resources were similar to those of other methods, we also considered the feasible operation region and the generation efficiency of the CHP resource. To apply these operation characteristics into the generation expansion planning problem modeled as a MILP, we linearized the cost function and constraints of the CHP resource. We compared 5-year planning results between the conventional optimization model applying constant heat-to-power ratio and generation efficiency of the CHP resource and the proposed optimization model. The models considered supplying peak and mean loads of the integrated energy system, retrieving remarkable differences among the resulting plans. In fact, heat and electrical energy resources were installed appropriately according to variations of loads using the proposed method. Although the overall generation efficiency of the CHP resource using the proposed model was lower than that using the conventional one, the total costs of the proposed model were lower than those of the conventional one. We expect that the proposed method will allow planners and operators of integrated energy systems to design and optimize their systems under conditions very similar to the actual operation of the CHP resource.

Author Contributions: All the authors contributed to this work. W.K. designed the study, performed the analysis and wrote the first draft of the paper. J.K. contributed to the conceptual approach and thoroughly revised the paper.

Acknowledgments: This work was supported by the Korea Institute of Energy Technology Evaluation and Planning (KETEP) and the Ministry of Trade, Industry & Energy (MOTIE) of the Republic of Korea (No. 20181210301380).

Conflicts of Interest: The authors declare no conflict of interest.

Nomenclature

Indices

y	Project year index, from $[1 : N_Y]$.
i	Electrical energy resource index, from $[1 : N_{ER}]$.
j	Heat energy resource index, from $[1 : N_{HR}]$.
ch	Combined heat and power resource index, from $[1 : N_{CHP}]$.
c	Candidate unit index, from $[1 : N_C]$.
be	Section index of electricity generation efficiency for CHP resource, from $[1 : N_{BE}]$.
bh	Section index of heat generation efficiency for CHP resource, from $[1 : N_{BH}]$.

Variables of electrical energy resources

$P_e^{i,t,y}$	Generation output of the electrical energy resource, i, for hour, t, is allocated in project year, y (MW).
$F_e^{i,t,y}$	Fuel usage of the electrical energy resource, i, for hour, t, is allocated in project year, y (MWh).
$E_{EES}^{t,y}$	Stored energy of EES for hour, t, in project year, y (MWh).
$P_{ch,e}^{t,y}$	Charging power of EES for hour, t, in project year, y (MW).
$\varphi_{ch,e}^{t,y}$	Variable for linearizing charging power constraints.
$\varphi_{dch,e}^{t,y}$	Variable for linearizing discharging power constraints.
$v_e^{i,c,y}$	Status of candidate generating unit, c, of electrical energy resource, i, in project year, y.
$v_{ch,e}^{t,y}$	Binary variable for selecting charging or discharging operation of the electrical energy storage.

Variables of heat energy resources

$P_h^{j,t,y}$	Generation output of the heat energy resource, j, for hour, t, is allocated in project year, y (MW).
$F_h^{j,t,y}$	Fuel usage of the heat energy resource, j, for hour, t, is allocated in project year, y (MWh).
$E_{TES}^{t,y}$	Stored energy of TES for hour, t, in project year, y (MWh).
$P_{ch,h}^{t,y}$	Charging power of TES for hour, t, in project year, y (MW).
$\varphi_{ch,h}^{t,y}$	Variable for linearizing charging power constraints.
$\varphi_{dch,h}^{t,y}$	Variable for linearizing discharging power constraints.
$v_h^{j,c,y}$	Status of candidate generating unit, c, of heat energy resource, j, in project year, y.
$v_{ch,h}^{t,y}$	Binary variable for selecting charging or discharging operation of the heat energy storage.

Variables of CHP resources

$P_{CHP,e}^{ch,t,y}$	Electricity generation output of CHP resource, ch, for hour, t, is allocated in project year, y (MW).

$P_{CHP,h}^{ch,t,y}$	Heat generation output of CHP resource, *ch*, for hour, *t*, is allocated in project year, *y* (MW).
$F_{CHP,e}^{ch,t,y}$	Fuel usage of electricity output of CHP resource, *ch*, for hour, *t*, is allocated in project year, *y* (MWh).
$F_{CHP,h}^{ch,t,y}$	Fuel usage of heat output of CHP resource, *ch*, for hour, *t*, is allocated in project year, *y* (MWh).
$\eta_{CHP}^{ch,t,y}$	Overall generation efficiency of CHP resource, *ch*, for hour, *t*, is allocated in project year, *y* (MWh).
$\varphi_{chp,e}^{ch,be,t,y}$	Variable for linearizing constraints of selecting efficiency section of electricity generation of CHP resource.
$\varphi_{chp,h}^{ch,bh,t,y}$	Variable for linearizing constraints of selecting efficiency section of heat generation of CHP resource.
$v_{CHP}^{ch,y}$	Status of CHP resource, *ch*, in project year, *y*.
$v_{CHP,eff,e}^{ch,be,t,y}$	Binary variable for selecting efficiency segment of electricity generation of CHP resource.
$v_{CHP,eff,h}^{ch,bh,t,y}$	Binary variable for selecting efficiency segment of heat generation of CHP resource.

Parameters

$C_e^{i,c}$	Capacity of candidate generating unit, *c*, of electrical energy resource, *i*.
$C_h^{j,c}$	Capacity of candidate generating unit, *c*, of heat energy resource, *j*.
CC_e^i	Capital cost of electrical energy resource, *i*.
CC_h^j	Capital cost of heat energy resource, *j*.
CC_{CHP}^{ch}	Capital cost of CHP resource, *ch*.
$FOMC_e^i$	Fixed operation and maintenance cost of electrical energy resource, *i*.
$FOMC_h^j$	Fixed operation and maintenance cost of heat energy resource, *j*.
$FOMC_{CHP}^{ch}$	Fixed operation and maintenance cost of CHP resource, *ch*.
FC_e^i	Fuel cost of electrical energy resource, *i*.
FC_e^j	Fuel cost of heat energy resource, *j*.
FC_{CHP}^{ch}	Fuel cost of CHP resource, *ch*.
$VOMC_e^i$	Variable operation and maintenance cost of electrical energy resource, *i*.
$VOMC_h^j$	Variable operation and maintenance cost of heat energy resource, *j*.
$VOMC_{CHP}^{ch}$	Variable operation and maintenance cost of CHP resource, *ch*.
LT_e^i	Lifetime of electrical energy resource, *i*.
LT_h^j	Lifetime of heat energy resource, *j*.
LT_{CHP}^{ch}	Lifetime of CHP resource, *ch*.
$P_{CHP,e}^{ch,op}$	Operating point of electricity generation, *op*, of CHP resource, *ch*.
$P_{CHP,h}^{ch,op}$	Operating point of heat generation, *op*, of CHP resource, *ch*.
$\eta_{CHP,e}^{ch,be}$	Electricity generation efficiency of section, *be*, of CHP resource, *ch*.
$\eta_{CHP,h}^{ch,bh}$	Heat generation efficiency of section, *bh*, of CHP resource, *ch*.
ρ_{EES}^i	Index of electrical energy storage in electrical energy resource, *i*.
ρ_{TES}^j	Index of thermal energy storage in heat energy resource, *j*.
$\gamma_{e,max}/\gamma_{e,min}$	Maximum/Minimum generation limit of electrical energy resource.
$\gamma_{h,max}/\gamma_{h,min}$	Maximum/Minimum generation limit of heat energy resource.
$SOC_{EES,max}/SOC_{EES,min}$	Maximum/Minimum SOC limit of electrical energy storage
$SOC_{TES,max}/SOC_{TES,min}$	Maximum/Minimum SOC limit of thermal energy storage
$\eta_{EES,dch}/\eta_{EES,ch}$	Discharging/Charging efficiency of electrical energy storage
$\eta_{TES,dch}/\eta_{TES,ch}$	Discharging/Charging efficiency of thermal energy storage
$\gamma_{EES,dch}/\gamma_{EES,ch}$	Cleared discharging/charging rate for electrical energy storage.
$\gamma_{TES,dch}/\gamma_{TES,ch}$	Cleared discharging/charging rate for thermal energy storage.
$\eta_{EES,eff}$	Turnaround efficiency for electrical energy storage
$\eta_{TES,eff}$	Turnaround efficiency for thermal energy storage
γ_d	Interest rate

Appendix A. Linearization of Nonlinear Constraints on Energy Storage

Equation (10) shows the nonlinear constraints during discharging of energy storage resources. The expressions on the right-hand side of these constraints can be divided into separate equations given by

$$
\begin{cases}
\varphi_{dch,e}^{t,y} \geq -\left(1 - v_{ch,e}^{t,y}\right) \cdot Z \\
\varphi_{dch,e}^{t,y} \leq \left(1 - v_{ch,e}^{t,y}\right) \cdot Z
\end{cases} \forall t, \forall y,
$$

$$
\begin{cases}
\varphi_{dch,h}^{t,y} \geq -\left(1 - v_{ch,h}^{t,y}\right) \cdot Z \\
\varphi_{dch,h}^{t,y} \leq \left(1 - v_{ch,h}^{t,y}\right) \cdot Z
\end{cases} \forall t, \forall y,
$$

(A1)

$$
\begin{cases}
\varphi_{dch,e}^{t,y} \geq \eta_{EES,dch} \cdot \sum_{i=1}^{N_{ER}} \left(\rho_{EES}^{i} \cdot \sum_{c=1}^{N_C} \left(C_e^{i,c} \cdot v_e^{i,c,y} \right) \right) - v_{ch,e}^{t,y} \cdot Z \\
\varphi_{dch,e}^{t,y} \leq \eta_{EES,dch} \cdot \sum_{i=1}^{N_{ER}} \left(\rho_{EES}^{i} \cdot \sum_{c=1}^{N_C} \left(C_e^{i,c} \cdot v_e^{i,c,y} \right) \right) + v_{ch,e}^{t,y} \cdot Z
\end{cases} \forall t, \forall y,
$$

$$
\begin{cases}
\varphi_{dch,h}^{t,y} \geq \eta_{TES,dch} \cdot \sum_{j=1}^{N_{HR}} \left(\rho_{TES}^{j} \cdot \sum_{c=1}^{N_C} \left(C_h^{j,c} \cdot v_h^{j,c,y} \right) \right) - v_{ch,h}^{t,y} \cdot Z \\
\varphi_{dch,h}^{t,y} \leq \eta_{TES,dch} \cdot \sum_{j=1}^{N_{HR}} \left(\rho_{TES}^{j} \cdot \sum_{c=1}^{N_C} \left(C_h^{j,c} \cdot v_h^{j,c,y} \right) \right) + v_{ch,h}^{t,y} \cdot Z
\end{cases} \forall t, \forall y.
$$

(A2)

These separate equations compose four inequality constraints for variable $\varphi_{dch,e}^{t,y}$ and $\varphi_{dch,h}^{t,y}$ and Z. For example, if either $v_{ch,e}^{t,y}$ or $v_{ch,h}^{t,y}$ is equal to 0 when during discharges, $\varphi_{dch,e}^{t,y}$ is the product of the discharging efficiency and the capacity of energy storage, as shown in Equation (A2). Otherwise, either $\varphi_{dch,e}^{t,y}$ or $\varphi_{dch,h}^{t,y}$ is equal to 0. Using the above inequality constraints and variables $\varphi_{dch,e}^{t,y}$ and $\varphi_{dch,h}^{t,y}$, the nonlinear constraints can linearized as follows:

$$
\sum_{i=1}^{N_{ER}} \left(\rho_{EES}^{i} \cdot P_e^{i,t,y} \right) \leq \varphi_{dch,e}^{t,y} \forall t, \forall y,
$$

$$
\sum_{j=1}^{N_{HR}} \left(\rho_{TES}^{j} \cdot P_h^{j,t,y} \right) \leq \varphi_{dch,h}^{t,y} \forall t, \forall y.
$$

(A3)

Likewise, the nonlinear constraints in Equation (11) can also be linearized as follows:

$$
\begin{cases}
\varphi_{ch,e}^{t,y} \geq -v_{ch,e}^{t,y} \cdot Z \\
\varphi_{ch,e}^{t,y} \leq v_{ch,e}^{t,y} \cdot Z
\end{cases} \forall t, \forall y,
$$

$$
\begin{cases}
\varphi_{ch,h}^{t,y} \geq -v_{ch,h}^{t,y} \times Z \\
\varphi_{ch,h}^{t,y} \leq v_{ch,h}^{t,y} \times Z
\end{cases} \forall t, \forall y,
$$

(A4)

$$
\begin{cases}
\varphi_{ch,e}^{t,y} \geq \eta_{EES,ch} \cdot \sum_{i=1}^{N_{ER}} \left(\rho_{EES}^{i} \cdot \sum_{c=1}^{N_C} \left(C_e^{i,c} \cdot v_e^{i,c,y} \right) \right) - \left(1 - v_{ch,e}^{t,y}\right) \cdot Z \\
\varphi_{ch,e}^{t,y} \leq \eta_{EES,ch} \cdot \sum_{i=1}^{N_{ER}} \left(\rho_{EES}^{i} \cdot \sum_{c=1}^{N_C} \left(C_e^{i,c} \cdot v_e^{i,c,y} \right) \right) + \left(1 + v_{ch,e}^{t,y}\right) \cdot Z
\end{cases} \forall t, \forall y,
$$

$$
\begin{cases}
\varphi_{ch,h}^{t,y} \geq \eta_{TES,ch} \cdot \sum_{j=1}^{N_{HR}} \rho_{TES}^{j} \cdot \left(\sum_{c=1}^{N_C} \left(C_h^{j,c} \cdot v_h^{j,c,y} \right) \right) - \left(1 - v_{ch,h}^{t,y}\right) \cdot Z \\
\varphi_{ch,h}^{t,y} \leq \eta_{TES,ch} \cdot \sum_{j=1}^{N_{HR}} \left(\rho_{TES}^{j} \cdot \sum_{c=1}^{N_C} \left(C_h^{j,c} \cdot v_h^{j,c,y} \right) \right) + \left(1 + v_{ch,h}^{t,y}\right) \cdot Z
\end{cases} \forall t, \forall y,
$$

(A5)

$$
P_{ch,e}^{t,y} \leq \varphi_{ch,e}^{t,y} \forall t, \forall y,
$$

$$
P_{ch,h}^{t,y} \leq \varphi_{ch,h}^{t,y} \forall t, \forall y.
$$

(A6)

Although the number of constraints increases, linearization allows to apply MILP-based optimization considering the charging mode and status of the energy storage resources.

Appendix B. Linearization of the CHP Fuel Consumption

In Equation (21), fuel consumption of the CHP resource should be linearized to realize MILP-based optimization because it is composed of the product of the binary variable for generation efficiency segment and the electricity generation. The linearization proceeds as follows:

$$
\begin{aligned}
F_{CHP,e}^{ch,t,y} &= \sum_{be=1}^{N_{be}} \left(\frac{P_{CHP,e}^{ch,t,y}}{\eta_{CHP,e}^{ch,be}} \cdot v_{CHP,eff,e}^{ch,be,t,y} \right) \\
&= \sum_{be=1}^{N_{be}} \left(g_{CHP,e}^{ch,be,t,y} \cdot v_{CHP,eff,e}^{ch,be,t,y} \right) \quad \forall ch, \forall t, \forall y, \\
F_{CHP,h}^{ch,t,y} &= \sum_{bh=1}^{N_{bh}} \left(\frac{P_{CHP,h}^{ch,t,y}}{\eta_{CHP,h}^{ch,bh}} \cdot v_{CHP,eff,h}^{ch,bh,t,y} \right) \\
&= \sum_{bh=1}^{N_{bh}} \left(g_{CHP,h}^{ch,bh,t,y} \cdot v_{CHP,eff,h}^{ch,bh,t,y} \right) \quad \forall ch, \forall t, \forall y,
\end{aligned}
\tag{A7}
$$

where either $g_{CHP,e}^{ch,be,t,y}$ or $g_{CHP,h}^{ch,bh,t,y}$ equals to the product of the generation output and the reciprocal of the segment generation efficiency. For simplicity during linearization, the product of variables is substituted by either $\varphi_{CHP,eff,e}^{ch,be,t,y}$ or $\varphi_{CHP,eff,h}^{ch,bh,t,y}$ as follows:

$$
\begin{aligned}
\varphi_{CHP,eff,e}^{ch,be,t,y} &= g_{CHP,eff,e}^{ch,be,t,y} v_{CHP,eff,e}^{ch,be,t,y} \forall ch, \forall be, \forall t, \forall y, \\
\varphi_{CHP,eff,h}^{ch,bh,t,y} &= g_{CHP,eff,h}^{ch,bh,t,y} v_{CHP,eff,h}^{ch,bh,t,y} \forall ch, \forall bh, \forall t, \forall y.
\end{aligned}
\tag{A8}
$$

These equations can be linearized based on linear inequality constraints as follows:

$$
\begin{cases}
\varphi_{CHP,eff,e}^{ch,be,t,y} \geq -v_{CHP,eff,e}^{ch,be,t,y} \cdot Z \\
\varphi_{CHP,eff,e}^{ch,be,t,y} \leq v_{CHP,eff,e}^{ch,be,t,y} \cdot Z
\end{cases} \forall ch, \forall be, \forall t, \forall y,
$$

$$
\begin{cases}
\varphi_{CHP,eff,h}^{ch,bh,t,y} \geq -v_{CHP,eff,h}^{ch,bh,t,y} \cdot Z \\
\varphi_{CHP,eff,h}^{ch,bh,t,y} \leq v_{CHP,eff,h}^{ch,bh,t,y} \cdot Z
\end{cases} \forall ch, \forall bh, \forall t, \forall y,
\tag{A9}
$$

$$
\begin{cases}
\varphi_{CHP,eff,e}^{ch,be,t,y} \geq g_{CHP,eff,e}^{ch,be,t,y} - \left(1 - v_{CHP,eff,e}^{ch,be,t,y} \right) \cdot Z \\
\varphi_{CHP,eff,e}^{ch,be,t,y} \leq g_{CHP,eff,e}^{ch,be,t,y} + \left(1 - v_{CHP,eff,e}^{ch,be,t,y} \right) \cdot Z
\end{cases} \forall ch, \forall be, \forall t, \forall y,
$$

$$
\begin{cases}
\varphi_{CHP,eff,h}^{ch,bh,t,y} \geq g_{CHP,eff,h}^{ch,bh,t,y} - \left(1 - v_{CHP,eff,h}^{ch,bh,t,y} \right) \cdot Z \\
\varphi_{CHP,eff,h}^{ch,bh,t,y} \leq g_{CHP,eff,h}^{ch,bh,t,y} + \left(1 - v_{CHP,eff,h}^{ch,bh,t,y} \right) \cdot Z
\end{cases} \forall ch, \forall bh, \forall t, \forall y.
\tag{A10}
$$

Finally, the fuel consumption of CHP is composed of the sum of single variables:

$$
\begin{aligned}
F_{CHP,e}^{ch,t,y} &= \sum_{be=1}^{N_{be}} \varphi_{CHP,eff,e}^{ch,be,t,y} \forall ch, \forall be, \forall t, \forall y, \\
F_{CHP,h}^{ch,t,y} &= \sum_{bh=1}^{N_{bh}} \varphi_{CHP,eff,h}^{ch,bh,t,y} \forall ch, \forall bh, \forall t, \forall y.
\end{aligned}
\tag{A11}
$$

Appendix C. Cost Data

Table A1. Cost data of energy resources

Resource Type	Unit Name	Overnight Capital Cost ($/MW)	Fixed O&M Cost ($/MW)	Fuel Cost ($/MWh)	Variable O&M Cost ($/MWh)	Life Span (Yr)	Candidate Size (MW)
Fuel-based Power Generator	DG1	900,000	15,000	33.2925	6.1	20	700, 600
	DG2	650,000	15,000	182.3	15	20	90, 80
Heat Only Boiler	HOB1	520,000	15,000	182.3	15	20	300, 250
CHP	CHP	1,150,000	5850	22.77	2.75	20	1200, 1000, 800
Electrical Energy Storage	EES	3,092,000	10,000	0	30	7	24, 20
Thermal Energy Storage	TES	3,184,000	12,000	0	30	7	30,20

References

1. Mavromatidis, G.; Orehounig, K.; Carmeliet, J. A review of uncertainty characterisation approaches for the optimal design of distributed energy systems. *Renew. Sustain. Energy Rev.* **2018**, *88*, 258–277. [CrossRef]
2. Koirala, B.; Chaves Ávila, J.; Gómez, T.; Hakvoort, R.; Herder, P. Local Alternative for Energy Supply: Performance Assessment of Integrated Community Energy Systems. *Energies* **2016**, *9*, 981–1004. [CrossRef]
3. Ashouri, A.; Fux, S.S.; Benz, M.J.; Guzzella, L. Optimal design and operation of building services using mixed-integer linear programming techniques. *Energy* **2013**, *59*, 365–376. [CrossRef]
4. Dominković, D.; Stark, G.; Hodge, B.-M.; Pedersen, A. Integrated Energy Planning with a High Share of Variable Renewable Energy Sources for a Caribbean Island. *Energies* **2018**, *11*, 2193–2207. [CrossRef]
5. Amusat, O.O.; Shearing, P.R.; Fraga, E.S. Optimal integrated energy systems design incorporating variable renewable energy sources. *Comput. Chem. Eng.* **2016**, *95*, 21–37. [CrossRef]
6. Amusat, O.O.; Shearing, P.R.; Fraga, E.S. Optimal design of hybrid energy systems incorporating stochastic renewable resources fluctuations. *J. Energy Storage* **2018**, *15*, 379–399. [CrossRef]
7. Fioriti, D.; Giglioli, R.; Poli, D.; Lutzemberger, G.; Micangeli, A.; Del Citto, R.; Perez-Arriaga, I.; Duenas-Martinez, P. Stochastic sizing of isolated rural mini-grids, including effects of fuel procurement and operational strategies. *Electr. Power Syst. Res.* **2018**, *160*, 419–428. [CrossRef]
8. Borelli, D.; Devia, F.; Lo Cascio, E.; Schenone, C.; Spoladore, A. Combined Production and Conversion of Energy in an Urban Integrated System. *Energies* **2016**, *9*, 817–833. [CrossRef]
9. Lo Cascio, E.; Borelli, D.; Devia, F.; Schenone, C. Future distributed generation: An operational multi-objective optimization model for integrated small scale urban electrical, thermal and gas grids. *Energy Convers. Manag.* **2017**, *143*, 348–359. [CrossRef]
10. Mathiesen, B.V.; Lund, H.; Connolly, D.; Wenzel, H.; Østergaard, P.A.; Möller, B.; Nielsen, S.; Ridjan, I.; Karnøe, P.; Sperling, K.; et al. Smart Energy Systems for coherent 100% renewable energy and transport solutions. *Appl. Energy* **2015**, *145*, 139–154. [CrossRef]
11. Yuan, R.; Ye, J.; Lei, J.; Li, T. Integrated Combined Heat and Power System Dispatch Considering Electrical and Thermal Energy Storage. *Energies* **2016**, *9*, 474–490. [CrossRef]
12. Geidl, M.; Koeppel, G.; Favre-Perrod, P.; Klockl, B.; Andersson, G.; Frohlich, K. Energy hubs for the future. *IEEE Power Energy Mag.* **2007**, *5*, 24–30. [CrossRef]
13. Dzobo, O.; Xia, X. Optimal operation of smart multi-energy hub systems incorporating energy hub coordination and demand response strategy. *J. Renew. Sustain. Energy* **2017**, *9*, 045501. [CrossRef]
14. Hemmati, S.; Ghaderi, S.F.; Ghazizadeh, M.S. Sustainable energy hub design under uncertainty using Benders decomposition method. *Energy* **2018**, *143*, 1029–1047. [CrossRef]
15. Dolatabadi, A.; Mohammadi-ivatloo, B.; Abapour, M.; Tohidi, S. Optimal Stochastic Design of Wind Integrated Energy Hub. *IEEE Trans. Ind. Inform.* **2017**, *13*, 2379–2388. [CrossRef]
16. Shahmohammadi, A.; Moradi-Dalvand, M.; Ghasemi, H.; Ghazizadeh, M.S. Optimal Design of Multicarrier Energy Systems Considering Reliability Constraints. *IEEE Trans. Power Deliv.* **2015**, *30*, 878–886. [CrossRef]
17. Wang, H.; Zhang, H.; Gu, C.; Li, F. Optimal design and operation of CHPs and energy hub with multi objectives for a local energy system. *Energy Procedia* **2017**, *142*, 1615–1621. [CrossRef]

18. Huang, H.; Liang, D.; Tong, Z. Integrated Energy Micro-Grid Planning Using Electricity, Heating and Cooling Demands. *Energies* **2018**, *11*, 2810–2829. [CrossRef]

19. Gambini, M.; Vellini, M. High Efficiency Cogeneration: Performance Assessment of Industrial Cogeneration Power Plants. *Energy Procedia* **2014**, *45*, 1255–1264. [CrossRef]

20. McDaniel, B.; Kosanovic, D. Modeling of combined heat and power plant performance with seasonal thermal energy storage. *J. Energy Storage* **2016**, *7*, 13–23. [CrossRef]

21. Pinel, P.; Cruickshank, C.A.; Beausoleil-Morrison, I.; Wills, A. A review of available methods for seasonal storage of solar thermal energy in residential applications. *Renew. Sustain. Energy Rev.* **2011**, *15*, 3341–3359. [CrossRef]

22. Li, J.; Laredj, A.; Tian, G. A Case Study of a CHP System and its Energy use Mapping. *Energy Procedia* **2017**, *105*, 1526–1531. [CrossRef]

23. Lahdelma, R.; Hakonen, H. An efficient linear programming algorithm for combined heat and power production. *Eur. J. Oper. Res.* **2003**, *148*, 141–151. [CrossRef]

24. Rong, A.; Lahdelma, R. CO_2 emissions trading planning in combined heat and power production via multi-period stochastic optimization. *Eur. J. Oper. Res.* **2007**, *176*, 1874–1895. [CrossRef]

25. Fang, T.; Lahdelma, R. Optimization of combined heat and power production with heat storage based on sliding time window method. *Appl. Energy* **2016**, *162*, 723–732. [CrossRef]

26. Kialashaki, Y. A linear programming optimization model for optimal operation strategy design and sizing of the CCHP systems. *Energy Effic.* **2018**, *11*, 225–238. [CrossRef]

27. Sheikhi, A.; Ranjbar, A.M.; Oraee, H. Financial analysis and optimal size and operation for a multicarrier energy system. *Energy Build.* **2012**, *48*, 71–78. [CrossRef]

28. Ko, W.; Park, J.-K.; Kim, M.-K.; Heo, J.-H. A Multi-Energy System Expansion Planning Method Using a Linearized Load-Energy Curve: A Case Study in South Korea. *Energies* **2017**, *10*, 1663–1686. [CrossRef]

29. Alipour, M.; Mohammadi-Ivatloo, B.; Zare, K. Stochastic risk-constrained short-term scheduling of industrial cogeneration systems in the presence of demand response programs. *Appl. Energy* **2014**, *136*, 393–404. [CrossRef]

30. Jiménez Navarro, J.P.; Kavvadias, K.C.; Quoilin, S.; Zucker, A. The joint effect of centralised cogeneration plants and thermal storage on the efficiency and cost of the power system. *Energy* **2018**, *149*, 535–549. [CrossRef]

31. Xie, D.; Lu, Y.; Sun, J.; Gu, C.; Li, G. Optimal Operation of a Combined Heat and Power System Considering Real-time Energy Prices. *IEEE Access* **2016**, *4*, 3005–3015. [CrossRef]

32. Korea Power Exchange (KPX). Load Forecast. Available online: http://www.kpx.or.kr/www/contents.do?key=223 (accessed on 6 November 2018).

33. Korea-District-Heating-Coorperation (KDHC). Heat and Electricity Business Status. Available online: http://www.kdhc.co.kr/content.do?sgrp=S23&siteCmsCd=CM3655&topCmsCd=CM3715&cmsCd=CM4487&pnum=10&cnum=81 (accessed on 6 November 2018).

34. Park, E.; Kwon, S.J. Solutions for optimizing renewable power generation systems at Kyung-Hee University's Global Campus, South Korea. *Renew. Sustain. Energy Rev.* **2016**, *58*, 439–449. [CrossRef]

35. Liu, Z.; Chen, Y.; Luo, Y.; Zhao, G.; Jin, X. Optimized Planning of Power Source Capacity in Microgrid, Considering Combinations of Energy Storage Devices. *Appl. Sci.* **2016**, *6*, 416–434. [CrossRef]

36. Lazard. Levelized Cost of Energy Analysis. Available online: http://www.lazard.com/perspective/levelized-cost-of-energy-analysis-100/ (accessed on 6 November 2018).

energies

MDPI

Article

Including Wind Power Generation in Brazil's Long-Term Optimization Model for Energy Planning

Paula Medina Maçaira [1],*, **Yasmin Monteiro Cyrillo [2]**, **Fernando Luiz Cyrino Oliveira [1]** and **Reinaldo Castro Souza [1]**

[1]   Industrial Engineering Department, PUC-Rio, Rio de Janeiro-RJ 22451-900, Brazil; cyrino@puc-rio.br (F.L.C.O.); reinaldo@puc-rio.br (R.C.S.)
[2]   Electrical Engineering Department, PUC-Rio, Rio de Janeiro-RJ 22451-900, Brazil; yasmin.cyrillo@engenharia.ufjf.br
*   Correspondence: paulamacaira@esp.puc-rio.br

Received: 31 December 2018; Accepted: 26 February 2019; Published: 2 March 2019

Abstract: In the past two decades, wind power's share of the energy mix has grown significantly in Brazil. However, nowadays planning electricity operation in Brazil basically involves evaluating the future conditions of energy supply from hydro and thermal sources over the planning horizon. In this context, wind power sources are not stochastically treated. This work applies an innovative approach that incorporates wind power generation in the Brazilian hydro-thermal dispatch using the analytical method of Frequency & Duration. The proposed approach is applied to Brazil's Northeast region, covering the planning period from July 2017 to December 2021, using the Markov chain Monte Carlo method to simulate wind power scenarios. The obtained results are more conservative than the one currently used by the National Electric System Operator, since the proposed approach forecasts 1.8% less wind generation, especially during peak periods, and 0.67% more thermal generation. This conservatism can reduce the chance of water reservoir depletion and, also an ineffective dispatch.

Keywords: net demand; wind power forecasting; long-term forecasting; intermittent sources; Markov chain Monte Carlo

---

1. Introduction

The depletion of traditional energy sources, such as water and fossil fuels, as well as concerns about the sustainability, safety and reliability of energy supply systems, has caused intense growth of wind power generation worldwide. In Brazil, in the last two decades wind energy share in the e energy mix has grown significantly, reaching 8.1% (13 GW) of installed capacity in July 2018, with 536 projects in operation [1]. Official expectations for 2026 indicate an increase of approximately 215%, reaching 28 GW of installed capacity [2] and another 229 enterprises in operation. Regarding the participation in the energy generation, in March 2006 wind power was responsible for only 0.075 GWh, 0.002% of the total generation, while in September 2017 this figure reached 5 TWh, representing 11% of the country's electricity generation that month.

Since Brazil is a country with continental dimensions, the Brazilian electricity sector (BES) is divided into four interconnected submarkets, corresponding to the country's geographic regions: North, Northeast, South and Southeast/Midwest. Figure 1 shows the high concentration (80%) of wind farms in the Northeast, a consequence of the presence of strong winds in the region and the complementarity between wind and water sources [3], in contrast to what happens in the Southeast/Midwest, which has only 0.55% of the projects. Most of the plants already under construction or planned for construction are also located in the Northeast, totaling 207 new ventures in the region of the 216 nationwide.

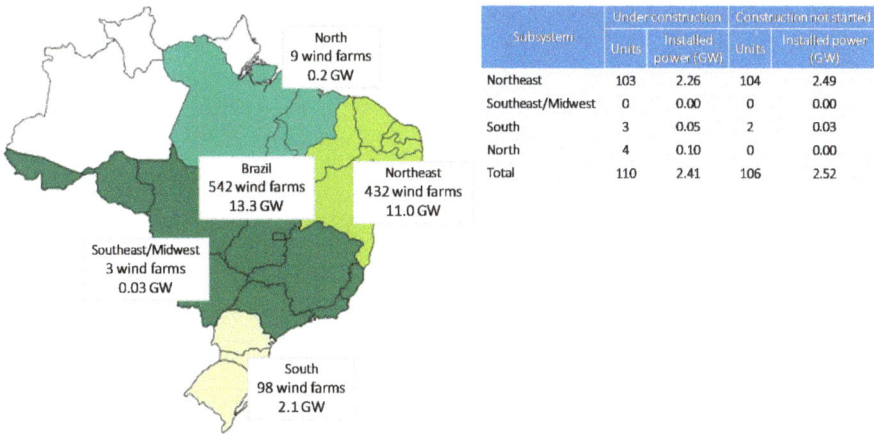

Subsystem	Under construction		Construction not started	
	Units	Installed power (GW)	Units	Installed power (GW)
Northeast	103	2.26	104	2.49
Southeast/Midwest	0	0.00	0	0.00
South	3	0.05	2	0.03
North	4	0.10	0	0.00
Total	110	2.41	106	2.52

Figure 1. Location of wind farms in Brazil and installed capacity figures [1].

Despite the increase of wind power generation, the Brazilian electrical system is basically formed by hydroelectric and thermal plants, which together represented approximately 70% of the country's total generation in 2017 [4–6]. Nowadays, planning the BES operation basically involves evaluating the future conditions of energy supply from hydro and thermal sources over the planning horizon to minimize the expected value of the operation cost during the planning period [7–9]. This cost is formed by the fuel costs plus penalties for failure of load supply, under operational and security constraints [10–12]. The other generation sources, including wind power, are discounted from energy demand in a deterministic way. That is, the current dispatch model does not consider wind generation's stochastic behavior. Suomalainen et al. [13] showed that in a system highly influenced by seasonal hydraulic generation, the high penetration of wind sources causes large impacts on the dispatch models. This is because wind generation mainly depends, on wind speed, a renewable and abundant resource, but one that is volatile. From an operational perspective, this volatility is a disadvantage, considering that for hydroelectric generation it is possible to minimize this variation through reservoir management [14].

In order to model the wind data at a particular site, an extensive historical series is necessary. If these datasets are not available, stochastic simulation techniques are required. The first important work dealing with simulation of wind speed data was published in 1996 and used statistical tests to check the adequacy of an autoregressive moving average (ARMA) model to provide this simulation [15]. Castino et al. [16] applied Markov chain and discrete autoregressive models to forecast wind series using information on wind speed and direction. Alexiadis et al. [17] predicted wind speed and power via artificial neural networks (ANN) and autoregressive models, showing that the first presents better accuracy. Papaefthymiou and Klöckl [18] presented a Markov chain Monte Carlo (MCMC) approach for the direct generation of synthetic time series of wind power output. Pinson [19] emphasized the importance of considering wind power generation as a stochastic process, and Zhang et al. [20] proved that wind power generation can be modeled as a stochastic process, since it is both nonlinear and unstable. Jung and Broadwater [21] presented a literature review of wind speed and power forecasting, involving spatial, probabilistic and offshore correlation, showing that the formulations cannot be compared since each model depends on the location (spatial correlation). Iversen et al. [22] proposed a modeling framework for wind speed based on stochastic differential equations. Landry et al. [23] described the probabilistic wind power forecasting method that was used to win the wind track of the Global Energy Forecasting Competition in 2014 (GEFCom2014). Aguilar et al. [24] developed a hybrid methodology using Singular Spectrum Analysis (SSA) and Conditional Kernel Density Estimation to achieve accurate probabilistic forecasts of wind output, and Cheng et al. [25] proposed an ensemble model for probabilistic wind speed forecasting. Ekström et al. [26] presented an improved method for the detailed statistical modeling of wind power generation based on a vector autoregressive model.

Besides wind generation, modeling and forecasting, this work aims to integrate these results in the current hydro-thermal dispatch model, without the need of any structural change in the optimization model, by updating the calculation of the non-dispatched plants. In order to reach the proposed objective and to consider the wind time series' stochastic behavior, the Frequency and Duration (F&D) methodological principle is used, combining via Markov chain Monte Carlo method the states of generation and load capacity to determine those ones of net demand and the corresponding probabilities. The F&D method is widely used to evaluate the static capacity adequacy for a given generation system, with the first application dating from 1968 [27]. This paper is based on the method developed by Leite da Silva et al. [28], which represents generation units by multistate models and load by hourly data. It also shows that not only the availability of the reserve states can be evaluated by discrete convolution, but also their frequencies. The methodology proposed here can be applied to other countries that face similar problems, that is, expanding reliance on intermittent sources while complementing the main sources, in addition to a centralized long-term operational planning without the possibility of modifying the equations of the optimization model, as, for example, Chile [29] and Nordic countries [30]. The Climate Forecast System Reanalysis combined with technical turbine information was used to obtain the historical time series from each wind farm, detailed in Section 3.1. All the statistical analysis and simulations were developed in R Statistical Software [31]. The optimization model was developed by the Stochastic Dual Dynamic Programming (SDDP) model [32]. In this sense, the novelty of this paper is the inclusion, via a stochastic approach, of the wind power generation into the optimal operation of the hydrothermal Brazilian system, resulting in the so called wind-hydro-thermal dispatch. Given that the wind power penetration is rather new in Brazil and is growing very fast, the lack of historical data is a real technical challenge for the implementation of such procedure.

The remainder of this paper is organized as follows: Section 2 presents a step-by-step framework to reach the goals described previously. Section 3 describes the results obtained, comparing our scenarios and the real datasets, as well as the wind power calibration outcome, and, finally the net demand calculation. Section 4 provides conclusions and future research directions.

2. Material and Methods

The introduction section emphasizes the growing trend of wind power generation in the Brazilian electricity mix and the need to consider the stochastic behavior of this source in the electric operation planning model. To meet this need, we propose a methodology that can be divided into three dependent procedures: historical measurements, Markov chain Monte Carlo (MCMC) modeling and the net demand approach.

Figure 2 presents a graphical framework with a description of all the calculations. In the first stage, each plant is associated with a wind speed history, obtained through measurements stations, and that speed is converted into wind power through the combination of turbine technical parameters. The data are calibrated from observed wind generation values. In the second stage, future scenarios of wind generation for each power plant are simulated using the Markov chain Monte Carlo method. In order to build the wind power generation of an entire submarket, the start-up date of each plant is considered. In the third stage, the expected energy load is considered with hourly frequency, being constructed through load profiles and future values. The calculation of the net demand is done through the combination of possible states of energy load and wind generation and probabilities are associated with each state. After obtaining the expected values of wind generation and net demand, it is necessary to include these data in the Brazilian hydro-thermal dispatch by replacing the wind energy values considered by the BES with the values obtained through the proposed approach. Such dispatch is based on an optimization model, where the income inflow is the stochastic variable. The objective function intends to minimize the sum of immediate and future cost of fuel and energy deficit, under operational and security constraints. The future cost is a convex combination of the expected value of the future cost function and the Conditional Value at Risk (CVaR) of that function, in order to insert risk aversion [33].

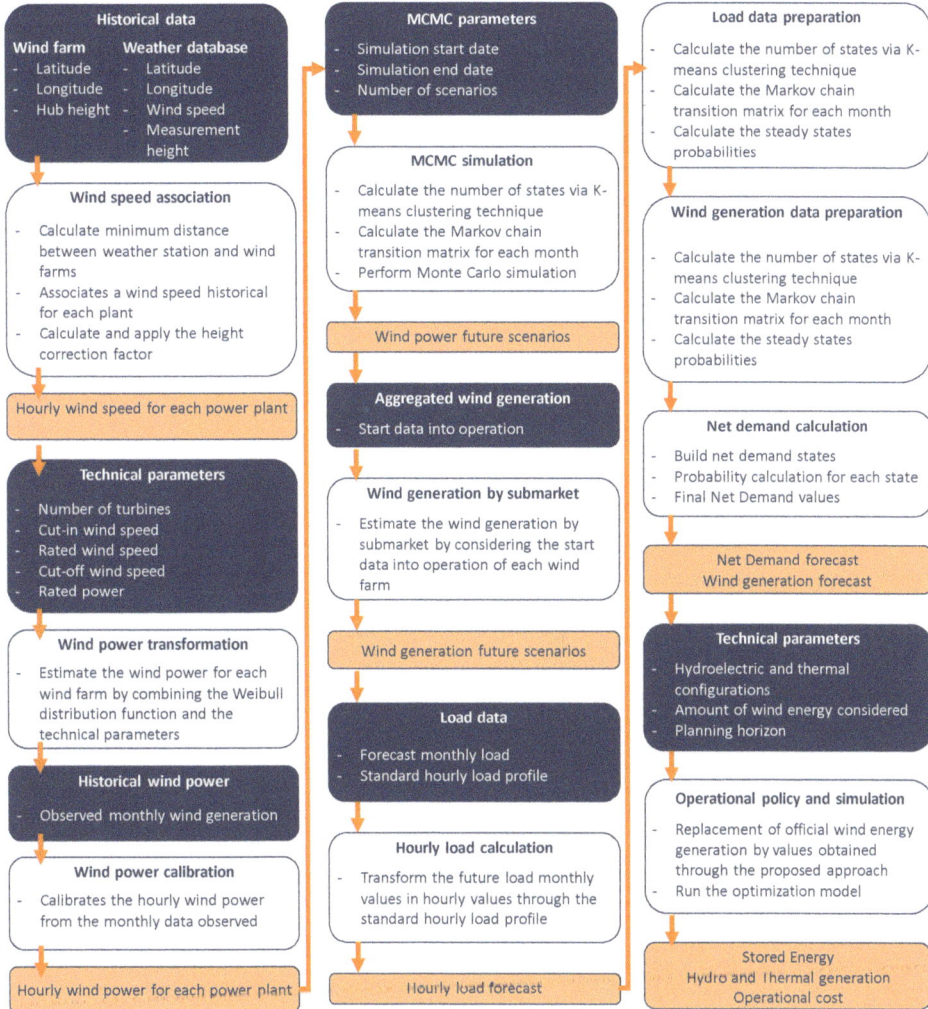

Historical data		MCMC parameters	Load data preparation
Wind farm	**Weather database**	- Simulation start date	- Calculate the number of states via K-means clustering technique
- Latitude	- Latitude	- Simulation end date	- Calculate the Markov chain transition matrix for each month
- Longitude	- Longitude	- Number of scenarios	- Calculate the steady states probabilities
- Hub height	- Wind speed		
	- Measurement height		

Historical data

Wind farm / **Weather database**
- Latitude / Latitude
- Longitude / Longitude
- Hub height / Wind speed
- Measurement height

MCMC parameters
- Simulation start date
- Simulation end date
- Number of scenarios

Load data preparation
- Calculate the number of states via K-means clustering technique
- Calculate the Markov chain transition matrix for each month
- Calculate the steady states probabilities

Wind speed association
- Calculate minimum distance between weather station and wind farms
- Associates a wind speed historical for each plant
- Calculate and apply the height correction factor

MCMC simulation
- Calculate the number of states via K-means clustering technique
- Calculate the Markov chain transition matrix for each month
- Perform Monte Carlo simulation

Wind generation data preparation
- Calculate the number of states via K-means clustering technique
- Calculate the Markov chain transition matrix for each month
- Calculate the steady states probabilities

Hourly wind speed for each power plant

Wind power future scenarios

Aggregated wind generation
- Start data into operation

Net demand calculation
- Build net demand states
- Probability calculation for each state
- Final Net Demand values

Technical parameters
- Number of turbines
- Cut-in wind speed
- Rated wind speed
- Cut-off wind speed
- Rated power

Wind generation by submarket
- Estimate the wind generation by submarket by considering the start data into operation of each wind farm

Net Demand forecast
Wind generation forecast

Wind power transformation
- Estimate the wind power for each wind farm by combining the Weibull distribution function and the technical parameters

Wind generation future scenarios

Technical parameters
- Hydroelectric and thermal configurations
- Amount of wind energy considered
- Planning horizon

Historical wind power
- Observed monthly wind generation

Load data
- Forecast monthly load
- Standard hourly load profile

Operational policy and simulation
- Replacement of official wind energy generation by values obtained through the proposed approach
- Run the optimization model

Wind power calibration
- Calibrates the hourly wind power from the monthly data observed

Hourly load calculation
- Transform the future load monthly values in hourly values through the standard hourly load profile

Stored Energy
Hydro and Thermal generation
Operational cost

Hourly wind power for each power plant

Hourly load forecast

Figure 2. Northeast region with the location of wind farms and CFSR stations.

It should be noted that there are other solutions for the inclusion of wind energy, and other so-called intermittent sources, directly in the dispatch optimization. Examples are the work of Papavasiliou et al. [34], who developed the multistage stochastic programming formulation; Jurasz et al. [35], who developed a mixed-integer nonlinear mathematical model; Morillo et al. [36], who included the expected production of wind energy in the objective function; and Raby et al. [29], who considered wind power as a new thermal plant. However, one of the main objectives of this work is to propose an approach that considers the variability of the wind series, but does not change the optimization model's formulation.

2.1. Historical Measurements

In the proposed methodology the first stage is related to the historical measurements, where the first step consists of obtaining a full year history of hourly wind speed series at the selected location. This information can be accessed through the Climate Forecast System Reanalysis (CFSR) [37] maintained by the National Centers for Environmental Prediction (NCEP). Through the CFSR it is possible to obtain the desired information according to geographic coordinates, with a spatial resolution of 0.25° by 0.25°. Thus, the association of a wind speed series to a wind farm was carried out by searching for the measurement point that minimizes the distance between them.

As a consequence of the difference in height between the measurement of wind speed (10 m) and the height of the turbines, it is necessary to consider a height correction factor [38], given that the greater the height the greater the wind speed generally it. This can be calculated by the natural logarithm ratio of the turbine height and the point at which wind measurement is performed, as in Equation (1):

$$HF_i = \frac{\log(HT_i)}{\log(HM_i)} \tag{1}$$

where HF_i is the height correction factor, HT_i is the turbine height and HM_i is the measurement height associated with wind farm i. So, the final wind speed, $WS_{h,d,m,i}$, associated with the wind farms is the result of multiplying the height correction factor by the original wind speed, $OWS_{h,d,m,i}$:

$$WS_{h,d,m,i} = HF_i \times OWS_{h,d,m,i} \tag{2}$$

where h is the hour, d is the day and m is the month.

According to Papaefthymiou and Klöckl [18], simulated wind power is more adequate than wind speed. There are two reasons for this (i) it avoids errors calculations in the conversion of speed to power; and (ii) the number of states is smaller, since the wind power is only observed for a given range of wind speed values. So, here we use the wind power series to simulate the future wind generation.

A convenient way to obtain the output power of a given wind turbine is through its power curve, which relates the resulting power of the turbine to a specific wind speed. Turbine manufacturers provide the power curves in tabular or graphic form. However, a generic equation that accurately represents this curve is needed in many problems involving wind power. The work of Kusiak et al. [39] identified that proper selection of power curve models is essential for predicting power and online monitoring of turbines accurately. Such models can be classified as discrete, deterministic/probabilistic, parametric/nonparametric and stochastic. Sohoni et al. [40] present a literature review of the existing methods for approximation of the power curve and the advantages and disadvantages involved in each one.

A parametric model defines the relation between input and output through a set of mathematical equations with a finite number of parameters. The transformation of wind speed into wind power ($WPA_{h,d,m,i}$) is made by a parametric model of the power curve of wind turbine expressed in Equation (3):

$$WPA_{h,d,m,i} = \begin{cases} 0 & WS_{h,d,m,i}\langle v_{ci}, \ WS_{h,d,m,i}\rangle v_{co} \\ q(v) & v_{ci} < WS_{h,d,m,i} < v_r \\ P_r & v_r \le WS_{h,d,m,i} \le v_{co} \end{cases} \tag{3}$$

where v_{ci} is the initial velocity at which wind starts generating power, and v_r is the cutting velocity, starting from which to the final velocity v_{co}, the power generated will be the same and equal to the rated power P_r of the turbine. For any velocity less than the initial one v_{ci} and greater than the final velocity v_{co}, there is no power generated. The relation between the resulting power and the wind speed between the initial velocity (v_{ci}) and the cutting velocity (v_r) is nonlinear and is represented by $q(v)$ in Equation (3). However, this relation can be approximated by different functions, polynomial or not. In this work, we selected the method based on the Weibull distribution (Equation (4)) [41,42], where the parameter k is the shape and is obtained by adjusting the maximum likelihood of the Weibull distribution function to the wind speed data from each month.

$$q(v) = \left(\frac{v_{ci}^{k_m}}{v_{ci}^{k_m} - v_r^{k_m}} + \frac{WS_{h,d,m,i}^{k_m}}{v_{co}^{k_m} - v_{ci}^{k_m}} \right) P_r \qquad (4)$$

So far, all the wind speed considered is converted into wind power ($WPA_{h,d,m,i}$), considering the turbine capacity factor. In this sense, the calibration year was the one when the data were obtained. That is, the calibration factor will be calculated from the observed monthly wind generation values and also the calculated monthly wind generation values, as in Equation 5, where $WP_{h,d,m,i}$ is the calibrated wind power, $WPA_{m,i}$ is the total wind power calculated for month m and wind farm i and $O_{m,i}$ is the observed wind generation at month m and wind farm i:

$$WP_{h,d,m,i} = WPA_{h,d,m,i} \times \frac{O_{m,i}}{WP_{m,i}} \qquad (5)$$

2.2. Stochastic Wind Power Simulation

The procedure developed in this study to stochastically simulate the wind power data is based on Almutairi et al. [43]. The first step involves the application of the K-means clustering technique [44] to transform the wind power data ($WP_{h,d,m,i}$) into a finite number of states ($WPS_{h,d,m,i}$). For the K-means technique, given a number k of clusters and M_k initial centroids, the distance $D_{k,c}$ from each wind power value WP_c to each M_c is calculated and all of the wind power values are assigned to the nearest centroid. New cluster centroids are calculated using the average of the wind power data in cluster k and the distances are recalculated. This process is repeated until the centroids remain fixed after a number of iterations. At the end, the calculated wind power values are replaced by the centroids of the cluster to which they belong, as explained in Equation (6), where c_k is the centroid value of cluster k:

$$WPS_{h,d,m,i} = \begin{cases} c_1, & \text{if } WP_{h,d,m,i} \in c_1 \\ c_2, & \text{if } WP_{h,d,m,i} \in c_2 \\ \vdots \\ c_k, & \text{if } WP_{h,d,m,i} \in c_k \end{cases} \qquad (6)$$

The second step corresponds to the creation of transition matrices, kxk, (P_{ind}), by month, where k is the number of states calculated in the previous step. The transition probability ($p_{a,b}$) from state a to b, for all indices $1 \leq (a,b) \leq k$, can be calculated by Equation (7):

$$p_{a,b} = \frac{n_{a,b}}{\sum_k n_{a,b}} \qquad (7)$$

where $n_{a,b}$ is the number of transitions from a to b. After obtain the transition probabilities for all each state is possible to construct the transition matrix for each month, as described in Equation (8):

$$P_{ind} = \begin{bmatrix} p_{1,1} & p_{1,2} & \cdots & p_{1,k} \\ p_{2,1} & p_{2,2} & \cdots & p_{2,k} \\ \vdots & \vdots & \ddots & \vdots \\ p_{k,1} & p_{k,2} & \cdots & p_{k,k} \end{bmatrix} \qquad (8)$$

Once the individual transition matrix P_{ind} based on individual state probabilities is obtained, the cumulative probability transition matrix P_{cum} can be created, so that its last column is equal to one for every row and month (Equation (9)):

$$P_{cum} = \begin{bmatrix} p_{1,1} & p_{1,1} + p_{1,2} & \cdots & p_{1,1} + p_{1,2} + \ldots + p_{1,k} \\ p_{2,1} & p_{2,1} + p_{2,2} & \cdots & p_{2,1} + p_{2,2} + \ldots + p_{2,k} \\ \vdots & \vdots & \ddots & \vdots \\ p_{k,1} & p_{k,1} + p_{k,2} & \cdots & p_{k,1} + p_{k,2} + \ldots + p_{k,k} \end{bmatrix} \tag{9}$$

The third and last step uses the cumulative Markov chain transition matrices and a uniform distribution to simulate the hourly wind power values for the T-years horizon $\left(WSIM_{h,d,m,y,i,s} \right)$. This procedure is described in detail in Almutairi et al. [43] and is summarized here. The initial state z is randomly selected, and a value between [0, 1] is generated from a uniform distribution; the next wind power state is determined by comparing the value of the random number with the elements of the zth row of the cumulative probability transition. If the randomly generated number is greater than the cumulative probability of the preceding state, but less than or equal to the cumulative probability of the succeeding state, the succeeding state is chosen to represent the next state. This procedure is repeated in order to simulate wind power data (by hour):

$$WSIM_{h,d,m,y,i,s} = c_k \quad if \quad c_{k-1} < p_{unif[0,1]} \leq c_k \tag{10}$$

where $WSIM_{h,d,m,y,i,s}$ is the simulated wind power for hour h, day d, month m, year y ($\sum y = T$), wind farm i and scenario s; c_k is the value of state k and $p_{unif[0,1]}$ is the value randomly generated from a uniform distribution.

2.3. Net Demand Calculation

The main task of this study is the inclusion of the wind power generation in the Brazilian hydro-thermal dispatch by considering its stochastic nature. For this purpose, it is crucial that all the required information for each wind farm be available. To create the wind generation data by submarket, it is first necessary to consider the starting date of operation in the wind power simulated in the previous step and then sum the wind farms corresponding to each submarket ($G_{h,d,m,y,j,s}$), as detailed in Equations (11) and (12):

$$G_{h,d,m,y,i,s} = \begin{cases} 0, & if \ d < d_{initial}, \ m < m_{initial}, \ y < y_{initial} \\ WSIM_{h,d,m,y,i,s} & otherwise \end{cases} \tag{11}$$

$$G_{h,d,m,y,j,s} = \sum_{i \in j} G_{h,d,m,y,i,s} \tag{12}$$

where, $d_{initial}$, $m_{initial}$ and $y_{initial}$ are the day, month and year of the starting date of operation, respectively; and j is the submarket.

To start the net demand calculation, the hourly load data are needed for the same horizon of the wind power generation. Due to the difficulty in finding official hourly load data in Brazil, a normalized standard load profile of the kind min-max ($LP_{m,h,j,t}$, where t is type of day) was created for the months, hours and type of day (weekdays, Saturdays and Sundays/holidays), and this is applied to the monthly load $ML_{m,y,j}$ (MW Average) expected by the government for each year of the horizon, resulting in the hourly load data, explained in Equation (13):

$$L_{h,d,m,y,j} = LP_{m,h,j,t} \times N_t \times ML_{m,y,j} \tag{13}$$

where N_t is the number of days according to the type of day that are in each month and year. After that, the K-means algorithm is used again to discretize both the wind power generation and load series into states ($LS_{h,d,m,y,j}$ and $GS_{h,d,m,y,j,s}$), see Equations (14) and (15), where c_k^L and c_k^G are the centroids value of cluster k for the load and wind generation time series:

$$LS_{h,d,m,y,j} = \begin{cases} c_1^L, & \text{if } L_{h,d,m,y,j} \in c_1^L \\ c_2^L, & \text{if } L_{h,d,m,y,j} \in c_2^L \\ \vdots & \\ c_k^L, & \text{if } L_{h,d,m,y,j} \in c_k^L \end{cases} \quad (14)$$

$$GS_{h,d,m,y,j,s} = \begin{cases} c_1^G, & \text{if } G_{h,d,m,y,j,s} \in c_1^G \\ c_2^G, & \text{if } G_{h,d,m,y,j,s} \in c_2^G \\ \vdots & \\ c_k^G, & \text{if } G_{h,d,m,y,j,s} \in c_k^G \end{cases} \quad (15)$$

In addition, in a subsequent step, the Markov chain transition matrices for each month (and scenario) are created, following the same steps described in Equation (8). The need to recalculate the states and transition matrices for the wind generation series comes from the fact that the series originated in the previous step has changed with the input of the starting dates of operation and the transformation into submarkets.

To be able to combine generation and load states in net demand states, it is necessary to associate a single probability to each state, and since in Markov chain theory the steady-state probabilities can be considered the long-term behavior of the system, after the effect of the initial conditions have decreased (i.e., in an equilibrium situation), this is the probability that represents the occurrence of each state. Then, the fourth step in this stage calculates the steady-state probabilities of each load and generation series for each month and year ($LSSP_{m,y,j}$ and $GSSP_{m,y,j,s}$).

According to Leite da Silva et al. [28], to find the model parameters of a combination $S = S(1) - S(2)$, let c_a and p_a denote the parameters associated with state a of component $S(1)$ and c_b and p_b be the parameters for state b of component $S(2)$. Suppose that the combination of states a and b gives state z of system S. Assuming that states a and b are statistically independent, then the parameters of S associated with state z are $c_z = c_a - c_b$ and $p_z = p_a \times p_b$. Since the net demand can be characterized as the difference between the load and the generation, the penultimate step of this procedure combines the load and generation model parameters to derive the net demand states and probabilities ($NDS_{m,y,j,s}$ and $NDSSP_{m,y,j,s}$) as in Equations (16)–(18):

$$NDS_{m,y,j,s} = LSSP_{m,y,j} - GSSP_{m,y,j,s} \quad (16)$$

$$P_{NDS}(NDS_{m,y,j,s}) = P_{LSSP}(NDS_{m,y,j,s} + GSSP_{m,y,j,s}) \times P_{GSSP}(GSSP_{m,y,j,s}) \quad (17)$$

$$NDSSP_{m,y,j,s} = NDS_{m,y,j,s} \times P_{NDS}(NDS_{m,y,j,s}) \quad (18)$$

The final output of this system is a value for each month and year in the forecasted period, and to do that the last step of the process is to calculate the expected values between the states and the probability associated with the net demand, resulting in a net demand value for each month and year, see Equation (19):

$$ND_{m,y,j} = \sum_s NDSSP_{m,y,j,s} \quad (19)$$

3. Results

In this section, the proposed methodology is applied to the Brazilian Northeast region in order to forecast the wind power generation from July 2017 to December 2021. The year 2016 is used as the base year, so the wind speed series extracted from CFSR are hourly from 1 January to 31 December (2016) and the standard load profile is built based on the hourly load for 2016, obtained from the National System Operator.

The Northeast region is composed by the states of Alagoas, Bahia, Ceará, Maranhão, Paraíba, Piauí, Pernambuco, Rio Grande do Norte and Sergipe, and has the longest coastline of the country's regions (3000 km). According to the National Electric Energy Agency (Agência Nacional de Energia Elétrica—ANEEL), in July 2017 there were 209 wind farms in operation, accounting for 3.6 GW of installed capacity, 172 under construction or due to start construction in the near future, totaling an addition of approximately 4 GW in this region in the following years. Of these 172 wind farms under construction, only 74 will be in operation within the study horizon, totaling 283 undertakings considered for all future calculations.

Figure 3 shows a map of the Northeast region and the location of wind farms. Note there is a concentration along the coast, but a considerable number are also located in the interior.

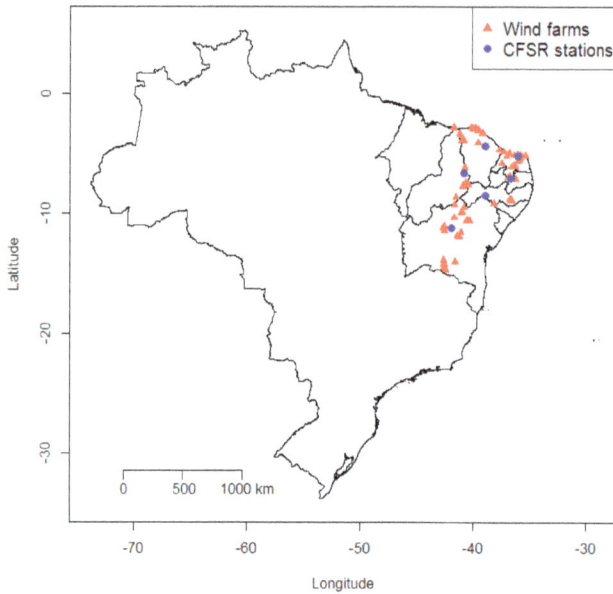

Figure 3. Northeast region with the location of wind farms and CFSR stations.

3.1. Obtaining the Historical Measurements Per Wind Farm

As mentioned above, the wind speed series were obtained from the CFSR system and the measured point to be used is the closest one to each wind farm. Following this, only 6 measurement points were needed and the maximum distance found was approximately 12 km. Table 1 shows the latitude and longitude of each measurement station of the CFSR system.

Table 1. Latitude and longitude of CFSR measurement stations.

Station	Latitude	Longitude
NASA 1362	−4.40	−39.07
NASA 1628	−5.21	−36.20
NASA 2046	−6.64	−40.91
NASA 2192	−7.05	−36.82
NASA 2622	−8.48	−39.07
NASA 3427	−11.14	−41.93

A survey of the turbine models installed at each wind farm revealed that only 7 turbine manufacturers serve this region with 12 models. Table 2 presents the manufactures, models and technical parameters.

Table 2. Manufacturers, models and technical information.

Manufacturer	Model	Hub Height (m)	Cut-in Wind Speed (m/s)	Rated Wind Speed (m/s)	Cut-Off Wind Speed (m/s)	Rated Power (kW)
Acciona	AW-3000/116	120	3	10.6	20	3000
Alstom	ECO-122/2700	139	3	10	25	2700
Alstom	ECO-86/1670	80	3	10	25	1670
Gamesa	G106/2500	93	2	12	24	2500
Gamesa	G97/2000	120	3	14	25	2000
GE	1.6-100	100	3.5	11	25	1600
GE	1.85-82.5	80	3	13	25	1850
GE	1.7-103	96	3	10	23	1700
GE	1.68-100	100	3.5	11	25	1680
Siemens	SWT-2.3-101	100	3	12	20	2300
Weg	AWG 110/2.1	120	2.5	11	20	2100
Wobben	E92/2350	138	2	13	25	2350

In this same step, the height correction factor is calculated, considering that the measurements are made 10 m above ground. The maximum wind turbine height found was 139 m and the minimum was 80 m, so the height correction factor varies from 1.90 to 2.14.

Therefore, by combining the Weibull distribution and the wind speed time series obtained from the CFSR system it was possible to obtain the expected wind power time series for each of the 283 wind farms. The year 2016 was used as calibration year, so for each wind farm, expected wind power is multiplied by a monthly calibration factor obtained through the relation between calculated and observed monthly generation. This calibration factor is calculated using monthly data since the observed wind power generation is available on a monthly basis. For the 74 farms that will be in operation, the monthly average of the calibration factor of the others plants was applied.

3.2. Wind Power Simulation Model Results

The first task in this stage is the transformation of the wind power obtained in the previous step into a finite number of states. To do this, we applied the K-means clustering technique where at least 98% of the data variability has to be represented, resulting in a variation from 13 to 24 in the number of clusters when applying the methodology to the wind power data, with great concentration around 15 clusters. The number of clusters varies accordingly to the original wind power data distribution, without any external interference.

The next step involves calculating the Markov chain transition matrix, constructed by the frequency at which one state transits to another, or to itself. As an example, in Table 3 the wind farm Abil transition matrix of March is presented. First, notice that the time series of this wind farm was divided into 14 finite numbers. Note also that the transitions happen in a gradual way, that is, if at a moment the generation is low, the chance that in the next moment this value is high is almost zero or zero in many cases, while the opposite is also true.

To build the cumulative transition matrix it is necessary to add for each state the probability of the previous states with their own probability. See Table 4 for the cumulative transition matrix for the Abil wind farm in March.

To simulate the hourly wind power values, it is first necessary to randomly select the initial state, for instance, state 3 (1.20 MW in the example) and then to choose a value from the uniform [0, 1] distribution. Assume it is 0.92. This means that the first simulated value is 1.20 MW and the second is 2.18 MW, since 0.92 is more than 0.72 (state 3) and less than 0.97 (state 4). This procedure continues until the entire horizon has been simulated for each wind farm. For convergence reasons, this entire process was repeated 200 times, thus generating the same number of possible scenarios. This value was chosen after performing sensitivity analysis of this database for different numbers of scenarios. Figure 4 depicts the average scenarios for the Abil farm.

Table 3. Transition matrix example.

State a (MW)	State b (MW)													
	0	0.32	1.20	2.18	3.32	4.47	5.71	7.03	8.57	10.20	11.77	13.55	15.53	17.60
0	0.50	0.50	0	0	0	0	0	0	0	0	0	0	0	0
0.32	0.01	0.84	0.13	0.02	0	0	0	0	0	0	0	0	0	0
1.20	0	0.14	0.58	0.25	0.03	0	0	0	0	0	0	0	0	0
2.18	0	0.04	0.21	0.41	0.28	0.04	0.03	0	0	0	0	0	0	0
3.32	0	0	0.05	0.20	0.34	0.23	0.06	0.08	0.02	0.02	0	0	0	0
4.47	0	0	0.02	0.07	0.22	0.31	0.24	0.10	0.03	0.00	0	0	0	0
5.71	0	0	0	0.04	0.07	0.21	0.29	0.21	0.09	0.07	0.02	0	0	0
7.03	0	0	0	0.03	0.03	0.10	0.24	0.32	0.15	0.06	0.05	0.02	0	0
8.57	0	0	0	0.00	0.02	0.06	0.04	0.21	0.31	0.17	0.08	0.08	0.04	0
10.20	0	0	0	0.00	0	0	0.04	0.09	0.26	0.24	0.26	0.07	0	0.04
11.77	0	0	0	0.02	0	0.02	0.02	0.05	0.11	0.30	0.23	0.11	0.11	0.02
13.55	0	0	0	0	0	0	0	0	0.08	0.04	0.31	0.31	0.15	0.12
15.53	0	0	0	0.04	0	0	0	0	0	0.11	0.19	0.11	0.30	0.26
17.60	0	0	0	0	0	0	0	0.03	0	0	0.02	0.03	0.14	0.78

Table 4. Cumulative transition matrix example.

State a (MW)	State b (MW)													
	0	0.32	1.20	2.18	3.32	4.47	5.71	7.03	8.57	10.20	11.77	13.55	15.53	17.60
0	0.50	1	1	1	1	1	1	1	1	1	1	1	1	1
0.32	0.01	0.85	0.98	1	1	1	1	1	1	1	1	1	1	1
1.20	0	0.14	0.72	0.97	1	1	1	1	1	1	1	1	1	1
2.18	0	0.04	0.25	0.66	0.93	0.97	1	1	1	1	1	1	1	1
3.32	0	0	0.05	0.25	0.59	0.83	0.89	0.97	0.98	1	1	1	1	1
4.47	0	0	0.02	0.09	0.31	0.62	0.86	0.97	1	1	1	1	1	1
5.71	0	0	0	0.04	0.11	0.32	0.61	0.82	0.91	0.98	1	1	1	1
7.03	0	0	0	0.03	0.06	0.16	0.40	0.73	0.87	0.94	0.98	1	1	1
8.57	0	0	0	0	0.02	0.08	0.12	0.33	0.63	0.81	0.88	0.96	1	1
10.20	0	0	0	0	0	0	0.04	0.13	0.39	0.63	0.89	0.96	0.96	1
11.77	0	0	0	0.02	0.02	0.05	0.07	0.11	0.23	0.52	0.75	0.86	0.98	1
13.55	0	0	0	0	0	0	0	0	0.08	0.12	0.42	0.73	0.88	1
15.53	0	0	0	0.04	0.04	0.04	0.04	0.04	0.04	0.15	0.33	0.44	0.74	1
17.60	0	0	0	0	0	0	0	0.03	0.03	0.03	0.05	0.08	0.22	1

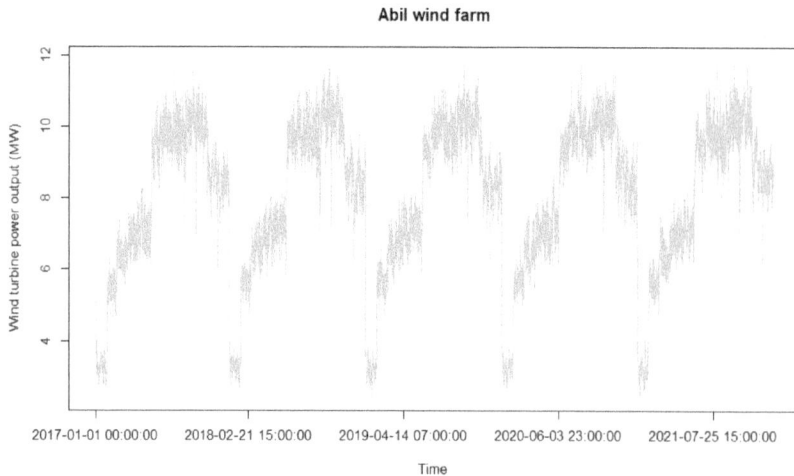

Figure 4. Abil average wind power scenarios.

In Figure 4 it is not possible to see the year-to-year variation of the wind series, but since the dispatch is optimized using the farms in aggregate form (submarket) rather than individually, the year-to-year variations are not significant, because in this case, the generation variation of a farm may be compensated for the others. Still, the year-to-year wind generation of the submarket is represented by considering the entry of new farms into operation in the future.

3.3. Checking of the Wind Power Simulation Model Results

In order to check the goodness of the simulated series, statistical factors were considered, such as the probability distribution function (PDF), autocorrelation functions (ACF), monthly characteristics and Wilcoxon signed-rank test.

The PDF shows the values of the distribution behavior of a given database. In this work it is used to compare the proportion of values generated through the simulation and the real series. Note in Figure 5 that the number of values belonging to a state in the simulated series is approximately equal to that of the real series, indicating that the synthetic scenario generation method used satisfactorily reproduces such behavior.

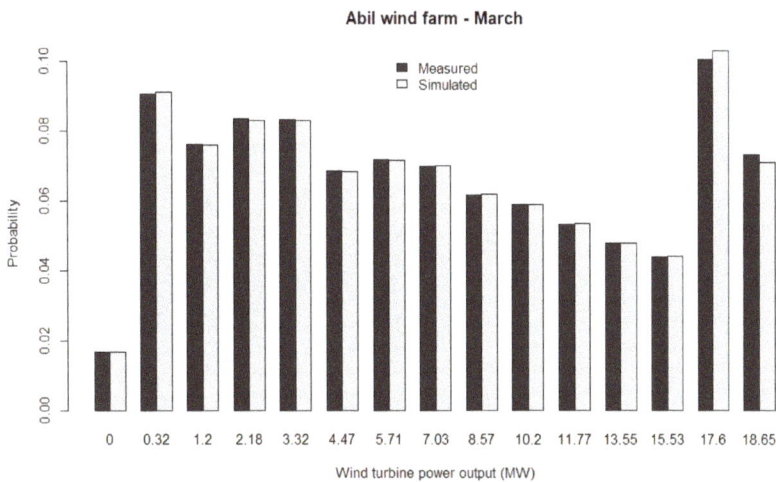

Figure 5. Comparison of PDFs.

Given that the wind time series behave differently between the seasons of a year, it is important to check whether the means and variations between the months are replicated for the simulated series compared to the measured series. Table 5 shows the mean and standard deviation values obtained from the measured data and the simulated data. Notice that the values of both, the simulated and observed series are very close, with the percentage errors varying from 0.08% to 3.64% for the mean and 0.00% to 1.26% for the standard deviation. This indicates that the scenario generation method replicates the inherent variation of the wind series.

In the previous analysis we checked whether the wind series seasonal behavior was replicated in the simulated series. In the present analysis, we check if the intra-hour behavior is reproduced using the autocorrelation function, which presents the correlation between the time intervals. Figure 6 shows that the peaks are exactly in the multiple of 24 lags, and since we are dealing with an hourly time series, it is possible to conclude that the present method also reproduces the hourly behavior.

Table 5. Monthly results.

Month	Mean (MW)		Standard Deviation (MW)	
	Measured Data	**Simulated Data**	**Measured Data**	**Simulated Data**
Jan	3.2204	3.2727	3.5331	3.5668
Feb	5.6109	5.5751	4.2978	4.3025
Mar	6.3998	6.4795	5.3218	5.3187
Apr	7.0073	6.9852	5.6116	5.6165
May	7.1462	7.1800	5.8127	5.8084
Jun	9.0496	9.3804	6.3166	6.3166
Jul	9.9104	9.9181	6.0918	6.1196
Aug	9.6687	9.7155	6.2547	6.2793
Sep	10.5886	10.2190	6.5827	6.6192
Oct	10.1114	10.2143	6.3178	6.3066
Nov	9.1185	8.7986	6.7229	6.6391
Dec	8.2729	8.3653	6.3252	6.3179

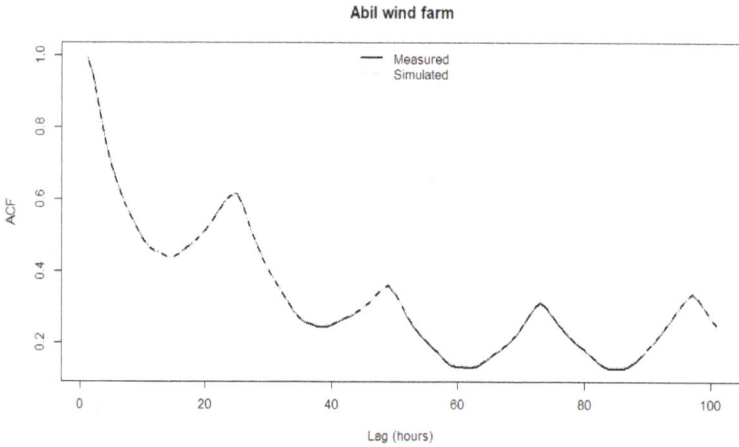

Figure 6. Comparison of ACFs.

The last feature to be tested is whether the simulated series fits the same distribution of the real data. In order to do that, the Wilcoxon signed-rank test was used (since the data were not normally distributed). The test's null hypothesis is that both distributions are equivalent, and since the p-value obtained is 0.2929, it is possible to conclude that the probability distribution of the simulated data is the same as the measured data.

3.4. Creating Hourly Load Values

As stated in Section 2.3, it is not possible to obtain hourly load forecasts directly from official organizations. Therefore, a methodology was developed to obtain such data from the load expectations made available for 60 months on monthly basis (Figure 7).

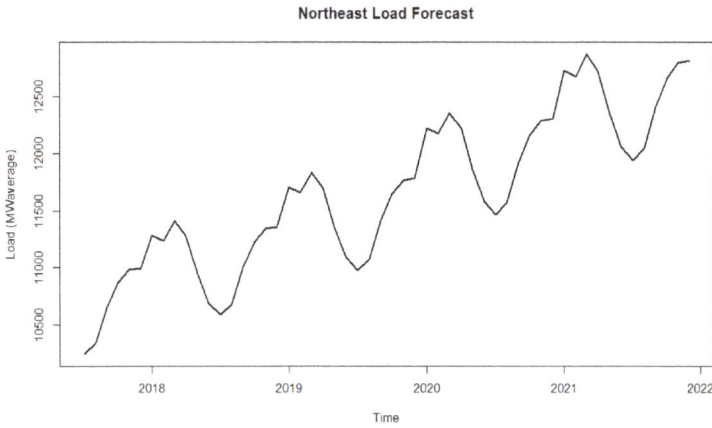

Figure 7. Northeast load forecast.

The first step involves obtaining the load profiles by type of day (workday, Saturday and Sunday/holidays) and month from a historical time series. Figure 8 presents the average monthly profiles for the Northeast submarket using data from January 2008 to December 2017. Note that during the early morning hours (1 a.m. to 5 a.m.) there is no significant difference in load between profiles, with the lowest values being on Sundays and holidays. However, during business hours (8 a.m. to 5 p.m.) there is a big difference between the profiles, where on weekdays the average load is much higher than on other days, as expected. In the evening (7 p.m. to 9 p.m.), this behavior is reversed, with higher load on Sundays and holidays.

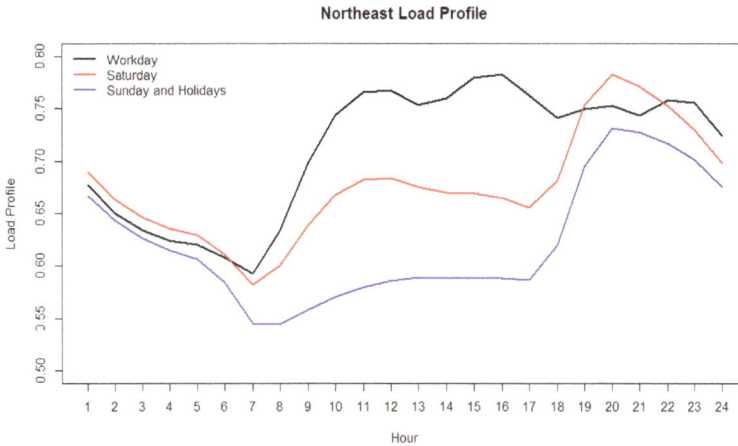

Figure 8. Northeast load profile by type of day.

By combining Figures 7 and 8, it is possible to transform the monthly forecast into hourly forecast, not displayed graphically for complexity reasons. The next step involves discretization of time series into finite values, just as was done for the wind power time series. The K-means process produced 17 clusters (Figure 9), accounting for 99.5% of the total data variation.

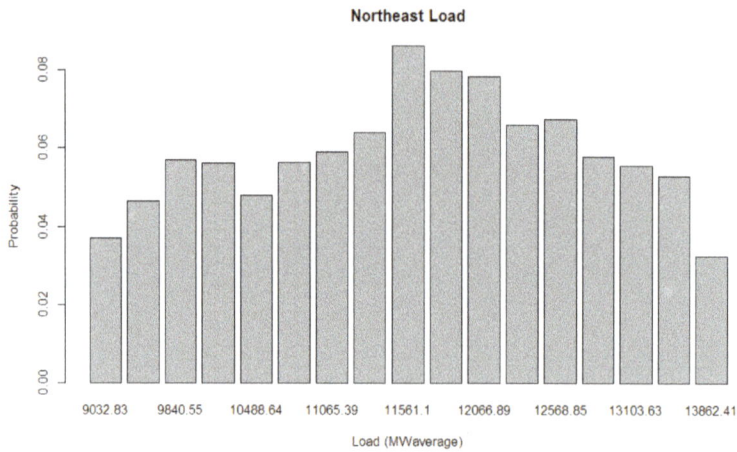

Figure 9. Northeast K-means load states.

3.5. Net Demand and Optimization of Dispatch Results

Now that all the information needed to calculate the net demand is available, it is possible to obtain the final results and compare then with those expected by the government.

The first and most important result is the wind generation forecast for the next 5 years (or 60 months). In Figure 10, the forecast through the net demand approach and from the official sources are presented, where it is possible to observe that, on average, wind power generation predicted by the government is higher than the simulated value, especially regarding peaks and valleys. Apart from this, both have the same trend and behavior.

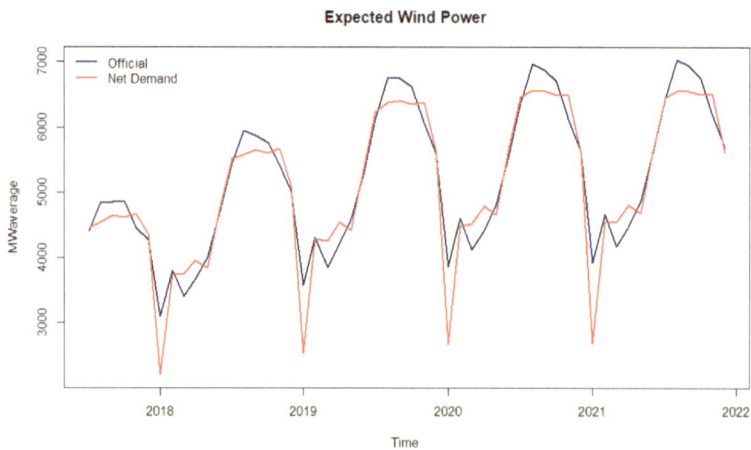

Figure 10. Northeast expected wind power.

As a consequence of a smaller expected wind power availability in the future compared to the official expectations, the load to be met (Figure 11) will present higher values when considering the methodology proposed here.

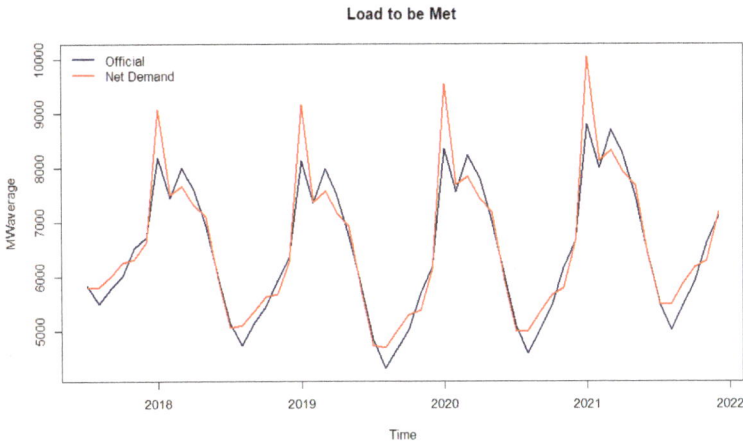

Figure 11. Northeast expected load to be met.

Considering the data from the load to be met (Figure 11), it is possible to evaluate, using the optimization dispatch model, the system's behavior according to the power demand forecasts and simulated wind power generation. Figure 12 presents the energy storage and Figure 13 gives the thermal generation according to official sources and the net demand approach. From Figure 12 it is possible to notice that the energy stored considering the official forecasts is almost the same as in the net demand approach, but in the dry season the proposed methodology is more conservative.

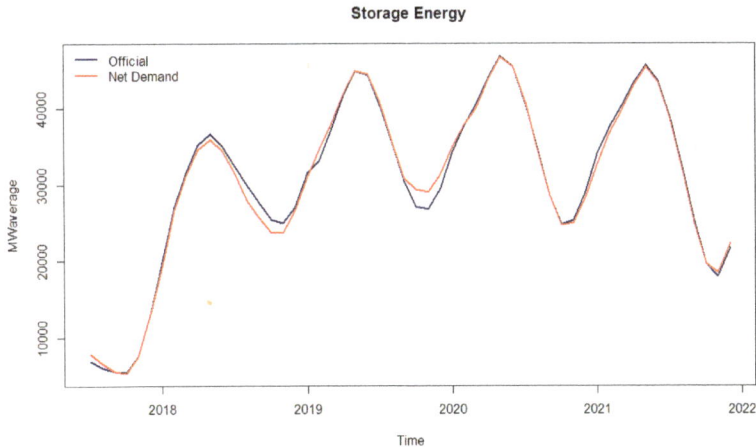

Figure 12. Northeast expected energy storage.

For the thermal generation, since with the net demand approach a more conservative generation of wind energy is expected, the complementation with thermal generation will be larger, as shown in Figure 13. Note that there was a drop in thermal dispatch in 2018, July 2018. In this month, it is possible to see that the load was low (the lowest level in 2018), and in addition, the income inflow contributed in a manner that it was not necessary to dispatch thermal power plants, since the hydro generation could supply most part of demand. It is important to emphasize that this is a part of the optimization results, restricted to the Northeast Brazil's region.

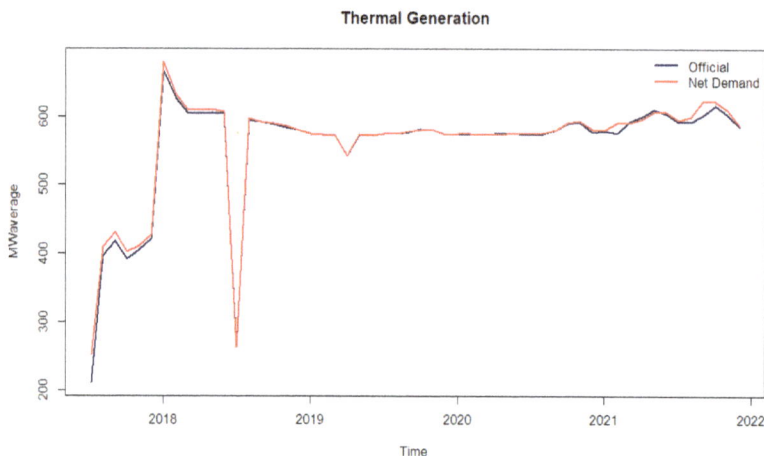

Figure 13. Northeast expected thermal generation.

4. Conclusions

This article proposes an approach to incorporate wind power generation in the current Brazilian hydro-thermal dispatch. The methodology is summarized in three main stages, comprising fourteen distinct steps, where the first stage deals with historical measurements, the second stage with the wind power simulation via MCMC and the final stage with the net demand calculation. The historical measurement stage involves obtaining wind speed data from a reanalysis database and transformation to wind power by the Weibull probability distribution. A wind power scenario, based on the MCMC model, is created for the entire time considering the starting date of operation of each wind farm. From statistical factors, the simulated time series characteristics are tested and validated in comparison with the observed series that replicate the monthly and hourly behavior and also the temporal correlation together with the probability distribution. In the net demand stage, a standard load profile by type of day is computed and applied to the monthly load forecasting, followed by the discretization of both the generation and the load series into finite states. In this stage, the Markov chain transition matrices and steady-state probabilities are also estimated. The combination of generation and load states into a single net demand value for each month of the planning period is performed by discrete convolution and expected value calculation.

The results obtained confirm that the expected wind power forecast using the proposed methodology is more conservative than the official expectations. That is, in periods of higher wind power generation in the Northeast, the net demand approach expects less generation than the government, with the same for periods with smaller generation. Thus, the load to be met will be greater according to the methodology proposed here. These values indicate a lower expectation of water storage in the future, translated into energy storage, and also higher generation from thermal plants. The main consequences of such differences between what is expected by the government and the forecast calculated here are the depletion of hydroelectric reservoirs and also the "non-optimization" of dispatch. Therefore, we can conclude that the consideration of probabilistic scenarios of wind energy generation as proposed in the net demand approach can mitigate errors in decision making by the Brazilian National Electric System Operator.

Finally, our objective was satisfactorily achieved, since the suggested methodology was able to: (i) reproduce the variable behavior of the wind series; and (ii) insert the wind generation in the current optimization dispatch model, conserving its structural mathematical formulation.

For future research, we suggest using other databases to provide wind speed series in addition to obtain a history of more than one year in order to capture yearly variability. We also suggest the application of other techniques to forecast wind power generation and consideration of different approaches for insertion of such generation in a mainly water and thermal dispatch. The application of methods that specifically consider the night/day effects of wind time series is another suggestion. We also recommend the construction of confidence intervals to increase the reliability of the applied simulation method, as was done in Cyrino Oliveira et al. [45] using bootstrap, or as it was done in Yang et al. [46] by applying Markov chain methods. Finally, it is possible to apply the proposed approach to other renewable sources, such as photovoltaic.

Author Contributions: Conceptualization, P.M.M., Y.M.C., F.L.C.O. and R.C.S.; methodology, P.M.M. and Y.M.C.; validation, P.M.M., Y.M.C., F.L.C.O. and R.C.S.; formal analysis, P.M.M. and Y.M.C.; writing—original draft preparation, P.M.M.; writing—review and editing, F.L.C.O. and R.C.S.; supervision, F.L.C.O. and R.C.S.

Funding: This study was financed in part by the Coordenação de Aperfeiçoamento de Pessoal de Nível Superior-Brasil (CAPES)-Finance Code 001. The authors also thank the R&D program of the Brazilian Electricity Regulatory Agency (ANEEL) for the financial support (P&D 0387-0315/2015) and the support of the National Council of Technological and Scientific Development (CNPq-304843/2016-4) and FAPERJ (202.673/2018).

Conflicts of Interest: The authors declare no conflict of interest.

References

1. ANEEL BIG–Banco de Informações de Geração [Generation Information Bank]. Available online: http://www2.aneel.gov.br/aplicacoes/capacidadebrasil/capacidadebrasil.cfm (accessed on 5 November 2017).
2. EPE & MME. *Plano Decenal de Expansão de Energia 2026*; EPE & MME: Rio de Janeiro, Brazil, 2017.
3. de Jong, P.; Sánchez, A.S.; Esguerre, K.; Kalid, R.A.; Torres, E.A. Solar and wind energy production in relation to the electricity load curve and hydroelectricity in the northeast region of Brazil. *Renew. Sustain. Energy Rev.* **2013**, *23*, 526–535. [CrossRef]
4. ONS Histórico da Operação–Geração de Energia. Available online: http://ons.org.br/ (accessed on 31 December 2018).
5. Huang, X.; Maçaira, P.M.; Hassani, H.; Cyrino Oliveira, F.L.; Dhesi, G. Hydrological Natural Inflow and Climate Variables: Time and Frequency Causality Analysis. *Phys. A Stat. Mech. Its Appl.* **2019**, *516*, 480–495. [CrossRef]
6. Maçaira, P.M.; Tavares Thomé, A.M.; Cyrino Oliveira, F.L.; Carvalho Ferrer, A.L. Time series analysis with explanatory variables: A systematic literature review. *Environ. Model. Softw.* **2018**, *107*, 199–209. [CrossRef]
7. Oliveira, F.L.C.; Souza, R.C.; Marcato, A.L.M. A time series model for building scenarios trees applied to stochastic optimisation. *Int. J. Electr. Power Energy Syst.* **2015**, *67*, 315–323. [CrossRef]
8. Ferreira, P.G.C.; Oliveira, F.L.C.; Souza, R.C. The stochastic effects on the Brazilian Electrical Sector. *Energy Econ.* **2015**, *49*, 328–335. [CrossRef]
9. Maçaira, P.M.; Oliveira, F.L.C.; Ferreira, P.G.C.; de Almeida, F.V.N.; Souza, R.C. Introducing a causal PAR(p) model to evaluate the influence of climate variables in reservoir inflows: A Brazilian case. *Pesqui. Oper.* **2017**, *37*, 107–128. [CrossRef]
10. Oliveira, F.L.C.; Souza, R.C. A new approach to identify the structural order of par (p) models. *Pesqui. Oper.* **2011**, *31*, 487–498. [CrossRef]
11. Souza, R.C.; Marcato, A.L.M.; Dias, B.H.; Oliveira, F.L.C. Optimal operation of hydrothermal systems with Hydrological Scenario Generation through Bootstrap and Periodic Autoregressive Models. *Eur. J. Oper. Res.* **2012**, *222*, 606–615. [CrossRef]
12. Duca, V.E.; Souza, R.C.; Ferreira, P.G.C.; Cyrino Oliveira, F.L. Simulation of time series using periodic gamma autoregressive models. *Int. Trans. Oper. Res.* **2019**. [CrossRef]
13. Suomalainen, K.; Pritchard, G.; Sharp, B.; Yuan, Z.; Zakeri, G. Correlation analysis on wind and hydro resources with electricity demand and prices in New Zealand. *Appl. Energy* **2015**, *137*, 445–462. [CrossRef]
14. Mendes, C.A.; Beluco, A.; Canales, F. Some important uncertainties related to climate change in projections for the Brazilian hydropower expansion in the Amazon. *Energy* **2017**, *141*, 123–138. [CrossRef]
15. Billinton, R.; Chen, H.; Ghajar, R. Time-series models for reliability evaluation of power systems including wind energy. *Microeletronic Reliab.* **1996**, *36*, 1253–1261. [CrossRef]
16. Castino, F.; Festa, R.; Ratto, C.F. Stochastic modelling of Wind velocities time series. *J. Wind Eng. Indutrial Aerodyn.* **1998**, *74–76*, 141–151. [CrossRef]

17. Alexiadis, M.C.; Dokopoulos, P.S.; Sahsamanoglou, H.S.; Manousaridis, I.M. Short-term forecasting os Wind speed and related electrical power. *Sol. Energy* **1998**, *83*, 61–68. [CrossRef]
18. Papaefthymiou, G.; Klöckl, B. MCMC for wind power simulation. *IEEE Trans. Energy Convers.* **2008**, *23*, 234–240. [CrossRef]
19. Pinson, P. Wind energy: Forecasting challenges for its operational management. *Stat. Sci.* **2013**, *28*, 564–585. [CrossRef]
20. Zhang, Y.; Wang, J.; Wang, X. Review on probabilistic forecasting of wind power generation. *Renew. Sustain. Energy Rev.* **2014**, *32*, 255–270. [CrossRef]
21. Jung, J.; Broadwater, R. Current status and future advances for Wind speed and power forecasting. *Renew. Sustain. Energy Rev.* **2014**, *31*, 762–777. [CrossRef]
22. Iversen, E.; Morales, J.; Mueller, J.; Madsen, H. Short-term probabilistic forecasting of wind speed using stochastic differential equations. *Int. J. Forecast.* **2016**, *32*, 981–990. [CrossRef]
23. Landry, M.; Erlinger, T.; Patschke, D.; Varrichio, C. Probabilistic gradient boosting machines for GEFCom2014 wind forecasting. *Int. J. Forecast.* **2016**, *32*, 1061–1066. [CrossRef]
24. Aguilar Vargas, S.; Castro Souza, R.; Pessanha, J.F.M.; Cyrino Oliveira, F.L. Hybrid methodology for modeling short-term wind power generation using conditional Kernel density estimation and singular spectrum analysis. *DYNA* **2017**, *84*, 145–154. [CrossRef]
25. Cheng, L.; Zang, H.; Ding, T.; Sun, R.; Wang, M.; Wei, Z.; Sun, G. Ensemble Recurrent Neural Network Based Probabilistic Wind Speed Forecasting Approach. *Energies* **2018**, *11*, 1958. [CrossRef]
26. Ekström, J.; Koivisto, M.; Ilkka, M.; Millar, R.J.; Lehtonen, M. A Statistical Modeling Methodology for Long-Term Wind Generation and Power Ramp Simulations in New Generation Locations. *Energies* **2018**, *11*, 2442. [CrossRef]
27. Hall, J.D.; Ringlee, R.J.; Wood, A.J. Frequency and duration methods for power system reliability calculations: Part 1–Generation system model. *IEEE Trans. Power Appar. Syst.* **1968**, *9*, 1787–1796. [CrossRef]
28. Leite da Silva, A.M.; Melo, A.C.G.; Cunha, S.H.F. Frequency and duration method for reliability evaluation of large-scale hydrothermal generating systems. *Gener. Transm. Distrib. IEE Proc. C* **1991**, *138*, 94–102. [CrossRef]
29. Raby, M.; Rios, S.; Jerardino, S.; Raineri, R. Hydrothermal system operation and transmission planning considering large wind farm connection. In Proceedings of the 2009 IEEE Bucharest PowerTech, Bucharest, Romania, 28 June–2 July 2009; Volume 1, pp. 1–8.
30. De Castro, C.M.B.; Marcato, A.L.M.; Souza, R.C.; Silva Junior, I.C.; Oliveira, F.L.C.; Pulinho, T. The generation of synthetic inflows via bootstrap to increase the energy efficiency of long-term hydrothermal dispatches. *Electr. Power Syst. Res.* **2015**, *124*, 33–46. [CrossRef]
31. *R Core Team R: A Language and Environment for Statistical Computing*; R Core Team R: Vienna, Austria, 2018.
32. Pereira, M.V.F.; Pinto, L.M.V.G. Multi-stage stochastic optimization applied to energy planning. *Math. Program.* **1991**, *52*, 359–375. [CrossRef]
33. Soares, M.P.; Street, A.; Valladão, D.M. On the solution variability reduction of Stochastic Dual Dynamic Programming applied to energy planning. *Eur. J. Oper. Res.* **2017**, *258*, 743–760. [CrossRef]
34. Papavasiliou, A.; Mou, Y.; Cambier, Y.; Scieur, D. Application of Stochastic Dual Dynamic Programming to the Real-Time Dispatch of Storage Under Renewable Supply Uncertainty. *IEEE Trans. Sustain. Energy* **2018**, *9*, 547–558. [CrossRef]
35. Jurasz, J.; Mikulik, J.; Krzywda, M.; Ciapala, B.; Janowski, M. Integrating a wind- and solar-powered hybrid to the power system by coupling it with a hydroelectric power station with pumping installation. *Energy* **2018**, *144*, 549–563. [CrossRef]
36. Morillo, J.; Pérez, J.; Zéphyr, L.; Anderson, C.; Cadena, A. Assessing the impact of wind variability on the long-term operation of a hydro-dominated system. In Proceedings of the 2017 IEEE PES Innovative Smart Grid Technologies Conference Europe (ISGT-Europe), Torino, Italy, 26–29 September 2017; Volume 1.
37. Saha, S.; Moorthi, S.; Wu, X.; Wang, J.; Nadiga, S.; Tripp, P.; Behringer, D.; Hou, Y.-T.; Chuang, H.; Iredell, M.; et al. NCEP Climate Forecast System Version 2 (CFSv2) Selected Hourly Time-Series Products. Available online: https://doi.org/10.5065/D6N877VB (accessed on 3 March 2017).
38. Manwell, J.F.; McGowan, J.G.; Rogers, A.L. *Wind Energy Explained: Theory, Design and Application*; Wiley: Hoboken, NJ, USA, 2010.

39. Kusiak, A.; Zheng, H.; Song, Z. On-line monitoring of power curves. *Renew. Energy* **2009**, *34*, 1487–1493. [CrossRef]

40. Sohoni, V.; Gupta, S.C.; Nema, R.K. A Critical Review on Wind Turbine Power Curve Modelling Techniques and Their Applications in Wind Based Energy Systems. *J. Energy* **2016**, *2016*, 1–18. [CrossRef]

41. Borowy, B.S.; Salameh, Z.M. Optimum photovoltaic array size for a hybrid wind/PV system. *IEEE Trans. Energy Convers.* **1994**, *9*, 482–488. [CrossRef]

42. Borowy, B.S.; Salameh, Z.M. Methodology for optimally sizing the combination of a battery bank and PV array in a wind/PV hybrid system. *IEEE Trans. Energy Convers.* **1996**, *11*, 367–375. [CrossRef]

43. Almutairi, A.; Hassan Ahmed, M.; Salama, M.M.A. Use of MCMC to incorporate a wind power model for the evaluation of generating capacity adequacy. *Electr. Power Syst. Res.* **2016**, *133*, 63–70. [CrossRef]

44. MacQueen, J. Some methods for classification and analysis of multivariateobservations. In Proceedings of the Fifth Berkeley Symposium on Mathematical Statistics and Probability, Oakland, CA, USA, 18–21 June 1967; p. 14.

45. Cyrino Oliveira, F.L.; Costa Ferreira, P.G.; Castro Souza, R. A parsimonious bootstrap method to model natural inflow energy series. *Math. Probl. Eng.* **2014**, *2014*, 1–10. [CrossRef]

46. Yang, X.; Ma, X.; Kang, N.; Maihemuti, M. Probability Interval Prediction of Wind Power Based on KDE Method with Rough Sets and Weighted Markov Chain. *IEEE Access* **2018**, 51556–51565. [CrossRef]

MDPI

St. Alban-Anlage 66

4052 Basel

Switzerland

Tel. +41 61 683 77 34

Fax +41 61 302 89 18

www.mdpi.com

Energies Editorial Office

E-mail: energies@mdpi.com

www.mdpi.com/journal/energies